THE TAO OF HOLISM

A blueprint for 21st century living

First published by O Books, 2008
O Books is an imprint of John Hunt Publishing Ltd., The Bothy, Deershot Lodge, Park Lane, Ropley, Hants, SO24 0BE, UK
office1@o-books.net
www.o-books.net

Distribution in:

UK and Europe
Orca Book Services
orders@orcabookservices.co.uk
Tel: 01202 665432 Fax: 01202 666219 Int. code (44)

USA and Canada
NBN
custserv@nbnbooks.com
Tel: 1 800 462 6420 Fax: 1 800 338 4550

Australia and New Zealand
Brumby Books
sales@brumbybooks.com.au
Tel: 61 3 9761 5535 Fax: 61 3 9761 7095

Far East (offices in Singapore, Thailand, Hong Kong, Taiwan)
Pansing Distribution Pte Ltd
kemal@pansing.com
Tel: 65 6319 9939 Fax: 65 6462 5761

South Africa
Alternative Books
altbook@peterhyde.co.za
Tel: 021 555 4027 Fax: 021 447 1430

Design: Stuart Davies
Cover design: Dr. David Jones

ISBN: 978 1 84694 080 4

A CIP catalogue record for this book is available from the British Library.

Printed by Chris Fowler International
www.chrisfowler.com

THE TAO OF HOLISM

A blueprint for 21st century living

DR HOWARD A. JONES

BOOKS

Winchester, UK
Washington, USA

CONTENTS

OTHER BOOKS BY THE SAME AUTHOR

Prescriptions for Health, William Collins, London, 1986

A Way of Health: Body, mind and spirit, Tynron Press, Stenhouse, Dunfriesshire, 1990

The Thoughtful Guide to God: Making sense of the world's biggest idea, O Books, 2006

PREFACE

Holism is a new science and a new way of thinking. Our past may be fragmented and our understanding narrow, but holism brings the strands together to give a coherent Oneness. These are essentially practical ideas that we need to start implementing now if we want our children and grandchildren to achieve peace, harmony and happiness in their lives. Although these are practical issues they are interwoven with a spiritual dimension, as our lives should be.

This commentary is based on the views of those who are experts in the various fields whose opinions most resonate with a holistic world-view. My aim is both to inform and to inspire others of the need for an awareness of the effects on other people and on the environment of everything that we do. There are many references at the end of this book to work that has informed or inspired me and I hope interested readers will follow up some of these subjects for themselves.

In order to apply ideas, it is essential to know how we have reached our current level of understanding. I think it is important to appreciate the history of ideas, so some biographical details of those involved in various aspects of holism, and background of the ideas themselves, are included in the main text or in supplementary Notes at the end of the book. As the popular saying goes: those who do not learn from history are destined to repeat it; or, to quote Coleridge, 'If men could learn from history, what lessons it might teach us! But passion and party blind our eyes, and the light which experience gives is a lantern on the stern, which shines only on the waves behind us!' References are given in square brackets in the text and are listed, together with other source material, at the end of the book.

Society as a whole will not change until we as individuals can persuade others by argument and by example. The *tao* means acting holistically to benefit ourselves, other people and the environment of which we are a part. The increasing violence, disrespect, insecurity and materialism of too many individuals in society cannot continue if we are to progress as a civilization and not decline into extinction.

Our physical health comes first, as without that we will not have the strength or vitality to do anything. The basic principles are discussed in Chapter 1. As our attitude of mind affects both our physical and spiritual health and the wellbeing of others, this is discussed in Chapter 2.

We need a breath of spiritual refreshment from art, music, literature or an appreciation of Nature so that we can develop our inner spiritual being as soul. This is our defining human quality and is how many express their spirituality. For others, formal religion represents the medium of their faith and the sustaining aesthetic dimension of their lives and this is the subject of Chapter 3. Our adherence to a particular religion may be purely a matter of custom, or indoctrination, or emotion, or faith, but whatever its origin, our spiritual beliefs should also be compatible with reason. The universe of science and of faith is one. We cannot put different aspects of our lives into different compartments, as if they belong to different people, or the same person living in two different worlds.

The Eastern religions tend to have a more holistic outlook than those of the West, so some ideas from Hinduism, Buddhism, Taoism and Confucianism are outlined in Chapter 3, together with the mystical divisions of the principal western religions of Judaism, Christianity and Islam – Kabbalah, Gnosticism and Sufism.

Myth and tradition portray allegorically the history of a tribe or society and form the foundations of religion through the scriptures. They play their part also in shaping the secular fabric of society, and this is the subject of Chapter 4. Myths connect us with our ancestral heritage as well as with the spirituality of the natural world.

In trying to construct for ourselves a path or way or *tao* to a holistic life we need to consider our role in society as well as those things that affect us as individuals. Chapter 5 is a survey of some sociological topics relevant to holism and it discusses those that sociologists feel are taking us in the wrong direction and how we might correct them.

The political and economic philosophy of western society comprises the subject matter of Chapter 6. It includes a sharp criticism by several writers of our continually expanding materialism and the devastating effects it is producing on us, physically and spiritually, and on the environment. Chapter 7 draws attention to the care we need to give our natural environment if it is to remain habitable into the next century. The education of our children is the focus of Chapter 8, for it is with them that the future of our planet and its civilizations lies.

It is not enough to attend to just one of these aspects of our life, though making a start anywhere is good and clearly we have to begin somewhere. But we need to embrace gradually all the dimensions of our existence and influence others to do the same to move to a state of wholeness within our society, the global community and the environment. Most of all, the philosophy of holism is about taking responsibility for our actions and seeing that these benefit others rather than neglecting or harming them.

This is a book with a message of hope overall. However, in parts, it is a chronicle of *impending* disasters, some of which have already begun, but which do not necessarily have to continue to develop. It is also a book of warnings – of how we must all individually and collectively take note of the overwhelming evidence that our materialist ways must be curbed. We must restore the spiritual and ecological consciousness of humankind. Earth is undergoing transformation largely as a result of human activity and we must transform ourselves. Our challenge now is to heal planet Earth and, with it, the social fabric of humankind.

I hope that this book will inform and inspire others to seek a holistic life-style, and that it will encourage some of the more egotistical power brokers to take a longer-term view and consider others yet unborn besides themselves.

INTRODUCTION

Indigenous tribes, where they have been left to live the kind of integrated lives they developed without interference from western capitalism or religion, live holistically. They could not have survived otherwise: those who did not live in harmony with their resources, perished. All the archeological evidence we have of early Man tells us that their societies had a great respect for the natural world and, of necessity, that they lived within the means of survival that were to hand. Little is known of the very earliest Chinese religious practices, but the earliest religion of Eastern civilization of which we have knowledge, amongst the people of the Indus, involved an all-encompassing spiritual deity and a unified view of all creation. This became the religion of Hinduism. Other religions subsequently developed in India and China – Buddhism, Taoism, Confucianism – that rejected the idea of any kind of remote divine agency having power over everyday human lives. But these religions retained three crucial elements of Hindu belief – individual responsibility for living a moral and productive life; the sanctity of all living beings; and the unity of all creation. The generally despotic rulers of their dynasties were treated like gods because of their power. All Eastern religion, theistic or atheistic, has always embraced a holistic vision of the universe, and all these faiths thrive to this day, with an increasing number of adherents in the countries of the West.

In the West, the evolution of humankind followed a different path. The ancient Egyptians, the Greeks and the Romans had a multiplicity of gods, generally serving pragmatic roles as guardians of the natural elements. The leaders of these civilizations were focused on more secular concerns, like the expansion of empires. Alongside and living within these empires there arose, around 2000 BCE, a tribe of peoples, the Hebrews or Israelites, who envisaged a single supreme deity having dominion over the lives of men. Their beliefs evolved into the religion of Judaism. An unbridgeable gulf existed between humankind and the God of the Hebrews, so this spiritual solitude was bridged by another group of people within the Roman Empire when they nominated a seer and healer called Jesus as a manifestation of God on Earth. The Roman emperor of the time, Constantine, was quick to seize on this as a means of exerting power and control over an empire of disparate pagan citizens. Having God here on Earth amongst humankind instead of as some remote and ill-defined being gave impetus to his authority. So the Christian religion was born through the human decision to designate the man Jesus as the Christ, the Messiah, the Redeemer of humankind. Those in control of manipulating followers of the faith did all they could to expunge records of most pagan ideas, but adapting those they saw fit to incorporate into the new faith. The pagan concept of a human being on Earth acting as a *Christos* or Saviour, Son of the Sky-God, had arisen many times before in other civilizations of both East and West. The pagan idea of a unifying natural cosmic spirituality was extinguished. Around six centuries after Jesus, the prophet Muhammad claimed insight into the will of this one deity of the Jews and his followers created the religion of Islam.

Because of the power of the Roman Empire, and then of subsequent followers of Christianity in the West, this religion has come to dominate the world faiths. About one-third of all those who profess a religion now claim to be adherents of Christianity. This religion has the most adherents not because of any intrinsic truth in its precepts but because of the zeal of its missionaries to Africa and the Americas and the ruthlessness with which it obtained 'converts' in Europe during the Crusades and Inquisitions. Subsequently, western Christian countries were empowered through the Renaissance of learning and the Industrial Revolution. The Catholic Church assumed authority over all matters, temporal and spiritual. It regarded itself as being able to provide all the answers to questions of both morality and the workings of the natural world. Such intolerant authority was bound to produce a backlash, and through the Reformation and Enlightenment, systems of rational thought and empirical investigation provided world-views different from those of the Catholic Church. After more than two millennia when the world order in the West was dominated by what were considered to be divine revelations, rationalism and empiricism were once again considered to be the source of knowledge of the world, as they had been in the time of the Greek philosophers. From early medieval times, theologians debated how the rational thought of the Greek philosophers could be reconciled with knowledge gained through revelation as presented in the scriptures.

The early Enlightenment rationalism of René Descartes (1596-1650) suggested that the physical or material things of the world, represented by the human body, were of an essentially different nature to things mental and spiritual, represented by the soul. Descartes' idea created the philosophy of dualism that has pervaded western thought ever since.

Science has made the progress that it has over the last four centuries by applying a principle, first suggested by Francis Bacon (1561-1626) in his book *The Advancement of Learning* (1605), that complex systems in Nature could best be studied and exploited for the benefit of humankind by examining each component separately, as if it were isolated from all that surrounded it. This is the philosophy of *reductionism* that has pervaded science – and much else in society – ever since. It has been immensely successful in helping us to understand the way the world works and our place in it – at least in a material sense. Science has given us the wonders of global technological communication, the benefits of advances in nutrition and medicine, and the enrichment of our lives in innumerable ways. But the knowledge gained by the new sciences came at a price. A division was created between what is human and what is non-human. The non-human was respected only insofar as it served the purposes of the human. With the ethos of Descartian dualism, scientific reductionism and the mathematical rationalism of Newton, any knowledge claimed by insight or intuition was increasingly viewed as unreliable, and only that obtained by reason or through the five senses was accepted as a true world-view. Our focus on rational minutiae and our dualistic philosophy have led to neglect of the unifying monistic outlook of holism.

Now we have come to the present age where attempts are being made to find common ground between the various religious faiths of the world, and between religion and science, so often regarded as having been in ideological conflict. With increasing international trade and with an awareness through television and the Internet of how other people live in parts of the world inaccessible to us as individuals, morality and commerce have taken on a global dimension rather than relating to just individuals or societies or nations. Man is a part of Nature and in destroying Nature we destroy ourselves. We have become too enchanted with the success of the rational, pragmatic, material philosophy of science and forgotten that there are other ways of looking at the world that have served humankind well for many millennia and still make a vital contribution to our physical and spiritual wellbeing. There is a dimension to our lives that we cannot live without if we are to find happiness and fulfillment, certainly as a global community, and that is our spirituality.

Although scientific reductionism has given us increased standards of living materially, we are becoming increasingly aware of the need for the spiritual dimension in our lives and awareness that so many facets of human existence are inter-related. Life on earth is intrinsically holistic, whether we acknowledge this fact or not. We have forgotten this in our admiration for the achievements of scientific rationalism. As American ecologist Thomas Berry said: 'we began to think of the universe as a collection of objects rather than as a communion of subjects' [27, p.16]. During the 20th century, and especially with the cultural revolution by young people during the 1960s, there was a strong movement to respect and explore the dynamic interaction between different physical and social systems, and especially the interaction between humankind and the natural environment, and between the materialist western outlook and the spiritual attitude of the East. This was perhaps the turning point that saw the re-emergence of the holistic outlook to living.

Over the past six or seven centuries, the western lifestyle has become increasingly powerful through a succession of cultural revolutions – the Renaissance, Reformation, Enlightenment, Industrial Revolution, Economic Revolution, Electronic Revolution – so that it now dominates the present existence of life on Earth. Experts in many different fields are now telling us that we must alter our behavioural mind-set if humankind is going to continue to exist on Earth. Our mental attitude to life and conscious awareness of our spirituality are all-important. Traditions of all kinds handed down within every culture from the ancestors have a role to play in this too. But religion needs to be seen for what it is – a man-made set of beliefs and practices inspired by the myths woven around the spiritual insight of one or two enlightened souls. Whether or not one regards this insight as divinely inspired is a matter of faith, but under no circumstances can any one world-view, secular or religious, be imposed by force on others if we want to live in harmony.

We cannot live on this planet heeding only our own selfish desires. We have to be aware of what others in our community, our nation, and increasingly in other nations, think and feel, so the social dimension is another aspect of holistic living we must embrace. Since the rise of science, with all that it has told us about the way the world works; the Industrial Revolution, which put many of these ideas into practice to improve the length and quality of our lives; and the Economic Revolution that established capitalism as the prevailing western economic system of day-to-day existence, the consequences of the application of this materialist philosophy have had devastating effects on our environment. Economics and environmental issues are intimately interwoven in our everyday lives, and the consequences of this interaction have to be considered in exploring and developing a holistic world-view.

Finally, the future of our planet is in the hands of our children and their descendents, provided we leave them a stable, coherent society and an environment that is fit to inhabit. The education of our children into these holistic values must surely be our prime concern.

Overall, in our holistic quest, we are seeking a Middle Way, a path or *tao* that either avoids extremes and finds a centrist path of moderation, or else embraces the best of the ideas of the polarized world-views in equal measure so that we can adopt and use them to our advantage in pursuing our ideals. The opposing poles of outlook can be represented by the descriptions shown in Table 1.

Reductionism	⟷	Holism
Scientific		Religious
Empirical		Mystical
Materialistic		Spiritual
Rational/logical		Intuitive
Planned/calculated		Spontaneous
Literal		Metaphorical
Allopathic		Naturopathic
Western		Eastern
Yang		Yin
Ego/superego		Id/collective unconscious
Individual		Social/global
Body		Soul/spirit
Brain		Mind
Certainty		Probability
Newtonian physics		Quantum physics
Binomial distribution		Gaussian outcomes
(only 2 answers – ‘yes’ or ‘no’)		(range of answers – ‘maybe’)

Table 1. Contrasting world-views to which we seek either an equitable mix of opposites or a harmonious Middle Way, a holistic *tao* between extremes

I have used the term ‘holism’ without explaining specifically what it means. *Holism* is the name given to the philosophical belief that the network of processes in the natural world, ourselves included, is an integrated whole at some fundamental spiritual level – that every part interacts with every other, and that the contrasting world-views represented in Table 1 all contribute to wholeness. At the most basic material level – that of the fundamental particles of matter – scientists have told us throughout the 20th century that material particles and energy constantly interchange and represent a holistic milieu in themselves.

The concept of holism within a societal context seems to have been used first by French sociologist Emile Durkheim (1858-1917), whom we shall meet again in Chapter 5, but the English word is attributed to the South African statesman Jan Christiaan Smuts (1870-1950). In his 1928 book *Holism and Evolution* he defined holism as 'The tendency in nature to form wholes that are greater than the sum of the parts through creative evolution'. I take the term 'creative evolution' to embrace a concept somewhat like that of 'animism' or 'panpsychism' – that no living system can ever be completely described by a physical description of its components, however detailed. There is an inherent, numinous, spiritual dimension in every aspect of the natural world, a spirituality that is uniquely active in the human organism. As the shaman Gordon MacLellan says: 'All that exists lives' [157, p.138].

Another way of expressing the concept of holism then is to say that the whole is always greater than the sum of its parts, however substantial or even complete we may consider the parts. This applies to our own lives, and to the network of the natural world, both organic and inorganic. Thus we may pursue some activity in our daily working life with zeal, enthusiasm and professional integrity, and consider that this is the only meaningful component of our lives. But even here, we need to look at our jobs in a broader context of why we are doing whatever it is that we do, and for whom, and is it really in either our best interests or those of others if we focus on just one pursuit with unswerving dedication. Should we be taking up valuable resources, including our own time, to commute to jobs far away, or is there a slower-paced activity we could find nearer home, which may not provide as many demanding professional challenges or as high a salary but gives us time for family and friends?

Environmentally, we have found to our cost on more than one occasion that when we have focused on some particular aspect of the complex web of the natural ecosystem and tried to 'correct' it by the introduction of a new species to combat one problem, we may eliminate that one but then create others more disastrous in its place. It is often the best solution to let Nature take whatever path she will, and let our own lives fall into a less strained pathway in harmony with natural ecosystems instead of trying to dominate and exploit them for our own materialist ends. We are finding now in global warming and turbulent weather patterns the devastating consequences of our disregard for holistic natural ecosystems.

The ancient Greek philosopher Heraclitus (*fl. ca.* 450 BCE) said that 'all is change': the world functions by maintaining a dynamic equilibrium between opposites, like those listed in Table 1. Whether we look at natural ecosystems or the cosmic dance of fundamental particles, everywhere we find this restless movement between opposite poles of thought or existence. The German philosopher G.W.F. Hegel (1770-1831) saw human social history evolving in a similar way with a constant movement between ideals: one pole representing a thesis, evolving into the opposite pole, the antithesis, which blend together to generate a Middle Way, the synthesis. The dominant tenor of each age Hegel called *Geist*. Somewhat over a century later the psychologist C.J. Jung described a comparable social spirituality as the *collective unconscious* that transmitted ideas from one society to another through space and time. In the 1980s, Cambridge biologist Rupert Sheldrake envisioned spiritual *morphic fields* functioning in the same way. *Geist*, the collective unconscious and morphic fields are manifestations of the holistic spiritual realm, which some would identify with World Soul or Communal Soul as spiritual deity.

The individual entities or units of activity that make up the holistic realm, but which may each form a system in themselves, American philosopher psychologist, Ken Wilber, describes as *holons* – whole entities that are parts of other wholes [390, p.40]. Reality in all domains, he says, is basically composed of holons – those features of our existence that we consider as topics or fields of enterprise complete in themselves, but which in every case interact with others to form a greater whole. The term 'holon' was introduced by Arthur Koestler in his book, *The Ghost In The Machine* [204, p.45]. It is a model component, complete and integral in itself, but which is also a component of a larger entity and interacting with other components at a similar level. Thus, our bodies are made up of interacting organs, the organs of tissues, the tissues of interacting cells, each of which is made of molecules, that have constituent atoms, that are made up of fundamental particles, that are made up of – energy. Working the other way, each of us is a part of a family, our family a unit of society, the collection of which forms a nation . . . and so it goes on. All of these units together form a *holarchy* – simply a hierarchy of entities regarded as holons. What does the holon concept contribute to a holistic world-view? It makes us realize that any system, any activity we think of as an isolated and complete entity is in fact part of a wholly interacting hierarchical chain that interacts with other holarchic complexities – like the subjects discussed in the chapters of this book. Holarchies survive through cooperation between the holons, not competition [54, p.84]. As poet John Donne said 'no man is an island'. Humankind is a holarchy, as is the natural world of which we are a part. David Bohm's 'implicate order', in which every entity is a microcosm of the whole of which it is a part, applies to the whole cosmos [31]. Bohm maintained that we cannot talk about 'cause and effect' in relation to any event because there are myriad causes (though some far removed in space and time) and multiple effects.

So the *Tao* or Way of holism embraces blending our material, day-to-day existence with an overriding spiritual outlook that determines the way we behave. Belief in a universal spiritual realm of which we are a part and with which we interact in every moment of our existence in everything we do is the fundamental basis of a holistic world-view. When authoritarian religion gave way to science in the Enlightenment as the fount of knowledge, the world-view moved gradually from trust in superstition and the use of intuition to belief that only empiricism and rationalism could provide truth. Science operates exclusively through reproducible experimental testing and logical mathematical interpretation, and it has been immensely successful in improving our quality of life over the past four centuries by doing so. But there is a realization now that the intuitive approach that has given us the natural remedies of folk medicine, the traditions of social communities, and the myths that form the scriptural bases of religion also have a vital part to play in human holistic existence. The intuitive spiritual outlook is complementary to the scientific approach and does not conflict with it or replace it. Holism demands that we accept both the rational and the spiritual as ways to make sense of our world.

Our quest for a holistic lifestyle will need to explore fields as diverse as psychology, theology, sociology, economics, environmental studies and education. The study of humankind living holistically in harmony with the natural environment has become so important that it has led to the creation of another separate scientific discipline – that of ecology. Modern Man has learnt much about the subject from the way of life of indigenous people, and from studying families of plants and animals living instinctively in a holistic way in the natural world. We must ensure that we learn the necessary lessons in time.

CHAPTER 1

THE BODY

General principles of physical health

The right to a basic standard of health for every individual is enshrined in United Nations policy. The United Nations Committee on Economical, Social and Cultural Rights states: "The right to health is closely related to and dependent upon the realization of other human rights, as contained in the International Bill of Rights, including the rights to food, housing, work, education, human dignity, life, non-discrimination, equality, the prohibition against torture, privacy, access to information, and the freedoms of association, assembly and movement. These and other rights and freedoms address integral components of the right to health." [373]. So national health services in the developed countries of the West have a legal obligation to see that their citizens are cared for, particularly in old age when their medical needs may be greater.

Richard Wilkinson found that the healthiest societies were not the richest but those in which social integration is highest and income is distributed most evenly [387]. The inequalities of earning power in Britain and America, as well as the availability of so much food of poor nutritional quality, do not lead to healthy nations, despite their material wealth. Far too many people in these countries eat too much food and of the wrong quality so that obesity and the diseases it causes have become significant health issues. Meanwhile, millions in the developing countries die of starvation. If people ate healthily they would not have the capacity to eat 'junk food'. Finding out about the most nutritious food takes a little effort, but no more than, say, planning a holiday. Holidays last for just a week or two; our diet will affect the quality of the whole of our lives.

As a fundamental part of our individual quest to achieve a holistic existence on Earth we try to achieve the best health that we can, given our genetic make-up. This outlook makes sense because few of us would be happy if we were continually plagued by illness. In the time of the ancient Greeks, the philosophy of Hedonism – the pursuit of immediate pleasures and the avoidance of pain – soon gave way to that of Epicureanism in which pleasures of today were selected more carefully to ensure that there were no adverse repercussions in the future. The holistic spiritual approach to conditions such as binge drinking, drug overdose and eating to excess is that they are disorders of the mind or spirit: they serve as compensation for the feeling of emptiness inside from loss of connection to one another and to the universal spirit. Since Britain introduced the possibility of 24-hour opening of drinking places, supposedly to reduce binge drinking, the incidence has increased by some 30%.

Alcohol or drug abuse presents an escape from reality, superficial hedonistic indulgence in pleasure heedless of the adverse consequences to self and others later. If these interpretations are true, all such excesses are cries for help in replacing voids in the lives of the individuals concerned. Of course, governments have a vested interest in promoting, or at least not actively reducing, the number of addicts to alcohol and tobacco products or gambling because the taxes swell the Exchequer. The British government went to war with China between 1839 and 1842 in order to maintain the lucrative opium trade. But promoting addiction is a short-sighted policy, for the illness it causes is far more costly and damaging to the economy than maintaining a healthy population.

The principles of physical health are relatively simple – eating nourishing foods and avoiding those that contribute to excessive weight gain. It's a principle some call 'sacred eating' – to eat only those foods that nourish the body, to avoid those that make you ill, or sluggish and tired, or overweight, and to stop when the body is sated. This also implies drinking only minimal quantities of alcohol and not using 'recreational' drugs that severely damage the functioning of the body. There is no such thing as 'recreational drugs': re-creation means creating anew, and the only things created by drugs like alcohol, tobacco products, heroin and cocaine are sickness and death. It's a fundamental principle of holistic health that we should treat the body as a temple. Naturopaths and spiritualists have said for many generations that illness has more to do with how we think and behave than dis-ease arising from our body being invaded by micro-organisms.

Maintaining an optimal weight is one of the first goals of healthy living. Energy intake in food must be balanced with energy expenditure through exercise. Of the three main food groups, proteins build tissue mass, carbohydrates and lipids provide energy, lipids most of all. Some carbohydrates are classed as 'insoluble' because they are scarcely broken down in the body. These comprise 'fibre' and nutritionists have known for decades that a high fibre intake (of about 30 grams each day) reduces the risk of heart disease, cancer, obesity and diabetes.

There is a strong temptation to eat what and when others are eating in order to be sociable, or as a compensation for some other need that is not fulfilled. Every body is individual. Some people can eat a large number of calories and burn them off quickly. For others, of slower metabolism, burning off calories and reducing weight is difficult; but maintaining one's weight at an optimal level is an essential part of remaining healthy. Our lifestyles are much more sedentary than a generation or two ago when exercise was a natural part of everyday life. Now we have to find opportunities to 'work-out' to relieve stress and burn off calories from our pressured, sedentary lifestyles. Exercise improves the oxygen supply to the brain, to help us think more clearly and stimulates the body to produce endorphins, the happy hormones. This is why indigenous peoples are also contented peoples, despite their lives of deprivation by western standards. Even today, those who live in less developed countries are surprised that we find it necessary to have cycling and rowing machines, located indoors, that take us nowhere!

The so-called body mass index (BMI) gives a good guide as to whether our weight is in the correct range, and this is the assessment that many doctors use[1]. The exercise component of holistic health depends heavily on age and state of health. When we are young and in good health we can benefit from fairly vigorous exercise that we can maintain into middle age so that we can still remain active in old age[2].

Regular exercise also helps us to breathe more deeply. The pressures of our modern lifestyle, with its fast pace and constant change, excessive information and greater bureaucracy, have led to a tendency in most people to breathe shallowly by expanding the chest. This is the way we breathe under stress in the body's 'fight or flight' mode. A more natural relaxed breathing needs to be cultivated. This is achieved by expanding and contracting the belly, like a baby does naturally when it sleeps. Deep, relaxed breathing is an important part of reducing stress, either as a regular, daily activity or as an essential element of contemplative meditation: becoming aware of the breath and then concentrating on inhalation and exhalation is how we enter a meditative state.

Our greater knowledge of nutrition and improvements in sanitation led to a great increase in longevity during the 20th century. Childhood mortality has also decreased. As a result, perhaps we are becoming more cavalier about our presence on the planet, taking life too much for granted whatever we do to our bodies. Treating our bodies with respect will encourage us to have more concern for the welfare of others and that of the planet. It's all about a change of attitude and taking responsibility for our own health and the welfare of others. We must also listen to what our bodies are telling us if we have adverse reactions, like headaches or indigestion, after eating certain foods. Food is the body's fuel. If we don't feel comfortable after a meal, this is our body's way of telling us that we've put in the wrong fuel!

There is a school of thought that we would benefit from eating foods that are within their natural ripening or harvesting season in our particular region or country rather than importing so many foods from overseas. We live within a particular energy field with which we resonate, and food grown within this field will be easier to assimilate. As we shall see later, there are many examples of indigenous peoples in other lands living in reasonable health on diets that we in the West would regard as inadequate; and migrants frequently find difficulty in accommodating to the diet of their new homeland.

Eating should also be a relaxing and, where possible, a social experience. Gentle background music is conducive to good digestion. Lying on the floor in front of the TV while eating, as some children seem to do, is not to be recommended. Mealtimes provide a good opportunity for family cohesion. Modern work pressures seem to have intruded even here into this once enjoyable family social experience.

Food additives

The use of food additives is not new: salt, sugar, vinegar, herbs and spices to flavour and preserve food were used even in biblical times at least 5000 years ago. However, the range of food additives in use today is essentially a late 20th century phenomenon and form part of a web of social changes. Our capitalist economics has increased the work pressures on men and women and given them less time for domestication, less time to cook, bake or prepare meals from basic ingredients and hence there is a greater need for ready meals. Most food today is therefore processed and all of it contains additives because the food is transported such long distances back and forth from source to consumer, via distribution centres, and meanwhile has to be preserved. There are now thousands of additives in use and every processed foodstuff has one or more in their ingredients.

Food additives are substances added to food for technological reasons and which are not normally present in the food in its natural state. Sometimes, substances are added to increase the nutrient value of a food or to replace those destroyed by processing, e.g. the B vitamins in breakfast cereals. Such added nutrients are not classed as 'food additives'; neither are contaminants like pesticides that come through the food production processes to reach the consumer; nor are the products of gamma radiation used to preserve food from bacterial contamination.

There are many different kinds of additives. Colourings are used to make food attractive in appearance as well as to distinguish different medications. Many colourings and some preservatives produce asthma or hypersensitivity in susceptible individuals and are thought to be responsible in some measure for the occurrence of ADHD in children [see 233, 234; 285][3].

Many additives are compounds found naturally in the body, for example, the anti-oxidants: vitamin C, is an anti-oxidant found naturally in fruit and vegetables, and used in fruit preserves; vitamin E, is an anti-oxidant found naturally in wheat germ and soya that is used extensively in fats in baking products. Almost all vitamins are lost or destroyed in food processing.

Flavourings and flavour enhancers are generally natural products, but used in concentrations higher than those to be found naturally. Artificial sweeteners are low-calorie synthetic compounds. Some have been found to be carcinogenic in animals at high concentrations[4]. For more details on additives, readers are referred to the books by Hanssen and Taylor [146; 362].

In assessing possible adverse effects of food additives (and drugs) we must remember that all these xenobiotics (compounds that don't belong in the body) are tested on animal models. Very many scientific and medical professionals have argued that it is not scientifically possible to reliably extrapolate findings from one species to another, so great are the different responses [13, 14]. The adverse effects of some of the colourings and preservatives described above however are not the results of animal tests: they have been found by their practical application in 'field trials' on the general human population! Furthermore, food additives are tested singly and there is almost no work at all on the effects of combinations of compounds. As the most popular junk food snacks contain at least five additives, the cumulative effect of all this biologically foreign material is unknown [178].

When we eventually return to a more holistic approach to life, as we will have to do as Earth's resources become depleted, the need for food additives will decrease. Food will be grown and eaten locally, thus reducing transportation to a minimum. The Department for the Environment says that the 'food miles' traveled have increased by 15% in Britain just over the past decade, which is damaging for the nutritional quality of the food and for the environment. If food is not simply a raw material for the food processing industry, there will be less need for preservatives, colourings and flavourings, and the food will be in a more natural and nutritious state when it arrives on our plates.

Diet and food quality

When we talk of dietary regimes we refer to those appropriate for the western population. It is interesting to note that different ethnic groups seem to have evolved with different dietary requirements that are suited to what the natural environment can provide locally. The Inuit of North America for example have a diet made up almost exclusively of meat and fish, with very little or no grain, fruit or vegetables, which are apparently essential nutrients for health amongst Westerners. Many adults in the Mediterranean countries seem to be grossly overweight, yet heart disease is rarely encountered. This is almost certainly because of the use of olives or olive oil in so many dishes. We shall see in Chapter 6 that the diet of the Ladakhis is very restricted but they have survived for many generations with none of the chronic diseases of the West.

A potential difficulty in establishing a healthy western diet arises with the quality of much of our food today. Those who study indigenous peoples living as they have done for many generations a life of ecological holism [42; 272; 305] find a surprisingly high standard of health or, at least, the absence of most of the West's degenerative diseases – arteriosclerosis (heart disease and stroke), cancer, diabetes, arthritis, Alzheimer's disease. They conclude that the cause of much disease in the West is our food – contamination of our food and water with fertilizers, drugs and pesticides, and the degradation of food by growing it in sterile soil, brought about by poisoning the soil and the animals with agrochemicals. This will be discussed further in Chapters 6 and 7, but the relevance here is to say that even if we eat fresh fruit and vegetables, there is no guarantee that they will contain the necessary amount of vitamins and minerals. This applies even to some organic food, for if it is grown on depleted soil, the produce, though free of pesticide residues, will still lack basic nutrients [156]. In some developing countries, where food intake has been increased through international aid, diseases linked to vitamin and mineral deficiencies persist or, in some cases, have even increased [329].

One of those who first raised concerns about the quality of our food was Weston A. Price (1870-1948), a dentist who ran a practice in Cleveland, Ohio, in the early years of the 20th century. He gave this up in the early 1930s to travel the world with his wife exploring the lifestyles of native peoples who had not yet had their ecologically sound holistic existence disrupted by the intrusion of western capitalism. He collected food samples from native diets and analyzed them in his laboratory. His findings, published in 1939 [305], showed levels of micronutrients four to ten times that of foodstuffs in the American diet of the time. His work inspired the nutritionist Sally Fallon to set up her own organization, the Weston A. Price Foundation in Washington, D.C., and she has updated Price's findings in a book of her own [95]. Price and Fallon's results are controversial because they are anti-vegetarian; they maintain that a diet that contains meat and milk from nutritionally fed cattle is beneficial to humans. As such foodstuffs are difficult to find today, there have been no extensive trials of their findings.

With the milk and dairy produce generally available in supermarkets, some authors are emphatically against including milk in the diet because contaminants are thought to predispose to cancer [56; 278; 291], and milk has long been implicated in producing many allergic or hypersensitivity reactions, particularly asthma and eczema [37; 264; 307]. A doctor, Walter Yellowlees, in clinical practice in Aberfeldy, in the Tay Valley in Scotland, reported the same deterioration in health of his rural patients when their ecologically sound natural lifestyle was replaced in the first half of the 20th century by the artificial agriculture imposed on them by government [400]. David Thomas has recently published his findings on the mineral depletion of foods grown in Britain since the beginning of World War II [366][5].

Natural agriculture is renewable and self-sustaining; industrial agriculture is destructive, exploitative and non-renewable. We are fast approaching a situation where natural agriculture will be our only option and David Pimental's work, referred to below, has shown that it is more economically viable. Industrial agriculture has only taken over in the last fifty years because of commercial pressures from our western capitalist economics. Even natural agriculture, however, will not sustain an indefinitely expanding world population.

Weston Price [305], Sally Fallon [95] and now Graham Harvey [156] are stressing quite correctly that the vitamin and mineral content of fruit and vegetables depend on the quality of soil where they are grown. The same is true for animal tissue, the nutrient quality of which depends on how they are fed. Whether or not you decide to be vegetarian is a matter of choice, for the nutritional science is equivocal. There is much to be said both for and against, though with the quality of meat generally obtainable today in supermarkets I would have little hesitation in recommending that intake of red meat be kept to the absolute minimum and preferably avoided completely. It is also more holistic and environmentally beneficial to be eating vegetable foods direct rather than feeding them to animals and then eating the animals [316, p.146]. Nuts, lentils and cereals are just as nutritious.

As Harvey correctly points out, 'plants rely heavily on microscopic bacteria and fungi to release essential minerals from rock particles' that make up the soil [156, p.23]. 'Without a large population of microbes, soils are physically unable to supply the plant with nutrients, even when chemical analysis may show them to be present' [*ibid.*]. The reason that the soil has become inert is that it has been poisoned by agrochemicals –

an array of fertilizers, pesticides and fungicides that kill bacteria and fungi, which would give life to the soil. Organic farmers use no chemicals: instead they rely on the decaying plant matter and recycled animal waste to re-nourish the soil which it will do continually. Agronomist David Pimental of Cornell University at Ithaca, New York, looked at the results of a 22-year long field trial carried out at the Rodale Institute of Pennsylvania [290]. In this trial, fields of corn and soya beans, two staple crops in America, were compared for productivity – one group was managed by organic crop rotation and the other group was kept alive only by pesticides and fertilizers. The organic fields over this time span produced more food, of better quality, and at lower cost. Fertilizers and pesticides in these quantities do not come cheaply, and there is the additional cost of the attempts to wash these toxic residues off the plants before human consumption, and to purify the water supply contaminated with the run-off into streams and rivers[6].

The amount of pesticide residue in any one food is tiny and is not likely to cause any ill effects in the short term, though pesticide residues are found more often than we would want[7]. However, as pathologist Vyvyan Howard has pointed out, we are ingesting these tiny amounts from many different sources every day, and we know nothing of their cumulative effects in the long term or the synergistic effects of a cocktail of all these chemicals [178]. The only way to avoid such contaminants is to eat organic food and drink fruit juices and spring water. Even there, we may not be getting an entirely pure product[8]. Living holistically, we can only eat food that is the purest we can obtain.

As Harvey says in respect of present-day food processing, farmers are no longer growing a food but producing a raw material for an industrial manufacturing operation [156, p.79]. Rifkin pointed out two decades ago that every step in that process consumes energy [316, p.157], and the situation has deteriorated since then. And Richard Manning points out: '70% of the grain crop of American agriculture goes to the livestock that replaced the bison that ate no grains, and one wonders, what is agriculture for?' [237].

Jeremy Rifkin is one of many authors encouraging consumption of as much food as possible in its native state rather than in processed form, and directly from vegetable sources rather than from reared animals [316, p.146]. Rifkin is not concerned in his book with the higher nutritional value of natural food but rather the waste of energy in transporting raw materials, manufacturing processed foodstuffs with addition of nutrients lost in processing, packaging, transporting to distribution warehouses, thence on to supermarkets, then to homes, and finally disposing of and recycling the packaging. It all involves a huge expenditure of energy. Livestock are fed largely on grain, and are then subjected to an equally long food processing chain. If we ate the grain itself in as near to its virgin state as possible, not only would we derive far higher nutrient value but we would save precious energy resources and reduce the production of greenhouse gases from the cattle. In the future, we will inevitably be moving more towards unprocessed vegetable food and eating less animal tissue and processed food as oil supplies run out in this century. Artificial farming is destroying the health of the world's population. It may produce a greater quantity of food, but of what quality and at what cost to the environment? The moral once again is that we need to eat and live holistically in harmony with our environment and keeping in mind what suits our particular metabolism: even health foods and supplements can cause problems in some people.

A comparable situation exists for livestock as for fruit and vegetables with regard to artificial farming. Instead of being allowed to graze on open mixed pasture comprising grass, clover, plantains and a host of meadow wild-flowers, too often cattle, in particular, are reared on a high-protein diet of cereals and soy bean meal (and, in the Britain of the 1980s, on ground-up animal carcasses, possibly including human remains [57]!) because these foodstocks are cheap and plentiful. The aim is that livestock put on more weight more quickly if they are beef cattle or produce more milk if they are part of a dairy herd. Soy and cereals are not part of the natural diet of cattle. In the U.S., they are also injected with growth hormones to the same end (though these are now banned in Europe) and with antibiotics (in the U.S and Britain) to kill infections like mastitis or tuberculosis. Organically produced meat and milk from cows fed on open pasture is higher in nutrients and lower in contaminants, some of which are thought to be responsible for the increases now seen in allergies in children and cancer and heart disease in adults [37; 56; 118; 291]. Between 1970 and 1990, the numbers of newly diagnosed cases of cancer in Britain increased by 45% for men and 55% for women [43]; this is unlikely to be due simply to better screening.

In the 1960s the average yield of a dairy cow in Britain was some 3500 litres of milk per year. Today, the average dairy cow will produce double this amount and, in the most productive herds, three times this volume of milk [156, p.126]. The pressure on farmers to be productive to these levels is such that most have gone out of business in Britain. In the early 1970s there were 100,000 farmers in Britain, but by April 2004, 84% of them had left farming [156, p.134]. During the decade of the 1990s, the price farmers received for their milk fell by a quarter. During the same period, the margins received by the major supermarkets almost doubled [254].

In America, milk production over the past decade has increased by some 25% while the dairy herd has reduced by about 10% in the same period [255]. This increased productivity has been achieved by the use of rBGH (genetically modified bovine growth hormone) manufactured by Monsanto. The use of this hormone is banned by the United Nations because it is known to produce increased incidence of cancer of the breast, colon and prostate [256].

Even grasslands are treated extensively today to produce a monoculture of rye grass instead of mixed flora. Supermarkets encourage farmers to produce only one or two crops so that they can grow the quantities the supermarket needs and they have to deal with only one supplier. Biodiversity is essential for the wellbeing of Earth's flora and fauna [396]. We are all familiar with the problems caused in humans by inbreeding, which is the principal reason for the prohibition of incest: the same applies to other species. Genetic modification of crops provides another potential danger. Despite precautions, open field trials simply cannot be contained as wind, birds and insects carry pollen far afield. There is already evidence that, when we eat genetically modified foodstuffs, the artificial genes alter the nature of the beneficial bacteria in the human gut [166; 268]. All these manipulations are undertaken ostensibly to feed the world but in practice their purpose is to swell the profits of the agrochemical industry.

Another damaging environmental impact is generated by the amount of plastic packaging used for food. To dispense with this seems admirable except for the fact that so many people use the toilet facilities in supermarkets then leave to handle their own and other people's food without washing their hands. So we cannot be surprised at the spread of diseases linked to bacteria like *E. coli* associated with faeces! As long as we have so many people who do not observe even the most fundamental hygiene it is best that food is wrapped.

Addiction

Although alcohol, in high enough concentration, may help to solubilize fat deposits from the arteries, the concentration needed to achieve this will do far more harm than good in its effect on the liver. There is no mention ever of alcohol or tobacco as a recommendation in a healthy holistic lifestyle. There are estimated to be four million alcoholics in Britain alone and 23,000 people died from alcohol-related diseases in 2005. In America, the number of alcoholics is estimated at 16 million and over one million in Australia [1]. Drinking red wine for its health benefits is not to be recommended. You can get the beneficial flavonoids in red wine from other sources and there is absolutely nothing good to be said about smoking, either for the smoker or for any of those who are unfortunate enough to be in his or her environment. Those who pledge to defy smoking bans are being selfish, inconsiderate and careless of their own health and everyone else's nearby. There is no moral reason whatever why people should be allowed to smoke in public places. This nonsense we hear about infringement of individual rights of smokers is as vacuous as if we tried to justify theft or rape on the same grounds. There are enough health problems arising from correctly prescribed drugs without adding to the tally with other drugs taken voluntarily.

Some health trusts now triage people with health problems who are known to be heavy smokers or consume way too much alcohol to the bottom of waiting lists. Ethically, in a situation of limited resources, a case can certainly be made for this if people become ill from their own over-indulgence and lack of self control. It is debatable whether it is acceptable that people who become ill through no fault of their own should be penalized because resources are insufficient to treat them, having been used up on trying to reverse effects of excess in irresponsible hedonists. Again, with limited resources at our disposal, more education in preventive medicine and individual responsibility would seem to be the most holistic and cost effective option to try to avoid this situation. This cannot be a matter for legislation; prohibition was tried without success in America. It has to be a matter of personal and social responsibility.

Perhaps a campaign could be mounted to make smokers and heavy drinkers aware that they are just as much drug addicts as those who inject themselves with cocaine or heroin. This needs to play a much more prominent role in education. The fact that not everyone who smokes dies of cancer, heart disease, stroke or emphysema does not disprove the connection between smoking and ill-health [193, p.285]. It is estimated that only about 5% of those who develop cancer or cardiovascular diseases can attribute their illness to heredity [393]. The other 95% have become ill because of imbalances in their lifestyles. The Mormons do not drink tea, coffee or alcohol and do not use tobacco products or drugs. As a result they have a much healthier lifestyle and are quite content without these stimulants. Such a lifestyle is clearly not intolerably stressful; on the contrary, supported undoubtedly by their faith, Mormons are generally quite tranquil people. Muslims and Baha'i are also prohibited from drinking alcohol, and Buddhists take no stimulating drugs of any kind, so these drugs are not necessary to provide happiness and, on the contrary, they do produce a great deal of misery for the addict as well as for those who are victims of their addiction.

Basically, the components of our lives – work, exercise, relaxation, recreation, sleep – need to be kept in balance in a healthy holistic lifestyle. Short-term imbalances present no problem – we may well encounter times when the work component, particularly, seems to take over; managers at work are demanding ever more productivity to maximize profit. But long-term imbalances are likely to cause problems in both physical and mental health. The pressures of modern living are encouraging us all to sacrifice sleep so that we can pack more 'achievement' into our 24/7 lives. It does no good to adults (or their children) for them to be addicted to work, however much money they earn; nor is it healthy for children to become addicted to computer games and videos instead of playing outdoors. For adolescents, gambling or playing slot machines in video arcades can become just as destructively addictive as smoking, alcohol or cocaine [137, 138].

When things go wrong with the body *

* The action of several alternative naturopathic preparations is described in this section, and the author has made every effort to ensure that such information is both accurate and up-to-date. Background orthodox medical or scientific research books and papers concerning the action of these compounds are given, where available. The author is a PhD in medical chemistry but is not a medical practitioner, orthodox or alternative. Therefore, any readers deciding to follow any of these suggestions must do so at their own risk and neither author nor publisher can accept responsibility for any effects, or lack of effectiveness, of any such preparations.

I have touched on some issues of ill-heath in the discussion so far but there are some topics we need to look at further.

A professor of psychology at the University of California, Theodore Roszak, believes that much of the physical and mental illness prevalent in society today can be traced to a loss of spiritual connection with our natural environment [196; 320]. He feels that many of society's ills could be reduced or even eliminated if we could re-establish this connection. He describes his study as ecopsychology.

Until we can regain this holistic health, in the short term we will have to rely on naturopathic and allopathic drugs. All medications are toxic to some extent, even if they have produced no ill effects in animal testing. Also, drugs are tested singly and the effects of various combinations of medication are virtually unknown [178]. Pharmaceutical drugs are almost all 'unnatural' compounds in that they are not normally found in the body (xenobiotics), and so have to be detoxified by the kidneys and liver. We simply have to balance the benefits from drugs, if they are really needed, with the potential extra stress to our body systems. All pharmaceutical preparations, or allopathic drugs as they are sometimes called[9], should be avoided unless they are really necessary to treat a chronic or life-threatening medical condition and when there is no simpler herbal or naturopathic remedy available that will produce the same result without side effects. We need to be especially wary of antibiotics because they kill helpful intestinal bacteria along with the toxic ones.

It is unfortunate that the medical profession is so firmly attached to treatment by allopathic drugs, dismissing all alternative treatments as quackery because they have not been verified 'scientifically'. In May of 2006, a group of physicians and scientists in Britain sent a letter to the chief executives of health trusts urging them to reject alternative medical treatments as 'unproven or disproved'. Their letter was reported in *The Times* newspaper [2]. In addition to being circulated within the NHS, the letter was also sent to Prince Charles, a known advocate of the use of alternative therapies. However, in orthodox 'scientific' medical practice there were thought to be some 40,000 errors in drug administration in British hospitals in 2005 alone. In America, there are approximately 7,000 deaths each year due to medication errors [251 (a)], and the FDA states that there is at least one death each day and 1.3 million people injured each year due to medication errors [251 (b)].

Many patients are also not given written information about the drugs they are prescribed. When we consider facts like these we must bear in mind that the medical profession is in thrall to the pharmaceutical industry and that anything they say will be biased as a result. Too often doctors ignore the fact that many complementary therapies are effective, and they rarely produce any adverse side-effects, in contrast to the plethora of adverse drug reactions of allopathic medicines. Adverse drug reactions remain at least the third or fourth biggest killer in the western world [13], and more recent studies suggest they may be *the* biggest killer [275]. These statistics take no account of the non-lethal distress caused by pharmaceutical drugs. Very often, patients seek help from a complementary therapist as a last resort because conventional medicine has no answers to their problems or the side effects of the prescribed pharmaceutical preparations are intolerable. The fallacy of regarding allopathic drugs as safe because they have been tested on animals was discussed above: humans often respond in a completely different way to animals, and frequently in a different way even from one another!

Alternative or complementary therapists describe their techniques as holistic because they deal with the whole body, physical and mental, as opposed to focusing on a particular organ or system, which is the norm in orthodox medicine. Alternative therapies (when used instead of allopathic drugs) or complementary therapies (if used in conjunction with orthodox medicine) are also 'active' in the sense that they require the participation of the patient in his or her treatment – adopting a positive frame of mind and making conscious decisions regarding treatment, often for a self-diagnosed condition. Orthodox medicine is more 'passive' in that the doctor makes the diagnosis and determines the prescription or treatment and the patient takes the tablets or follows the instructions unquestioningly; and pharmaceutical drugs will probably do their work whatever the mental attitude of the patient. Also, doctors are often insufficiently sensitive to the effects on the patient of informing them of their diagnosis, especially if they do not add some words of encouragement. The need to rethink our whole approach to what has become orthodox medicine in the West was pointed up by Brian Inglis a quarter of a century ago now [185], but his comments are still relevant. The situation he highlighted has, if anything, deteriorated since then because the range of allopathic drugs with their adverse side effects has increased considerably. Ivan Illich [183] and Lynne McTaggart [248] have shown us that we should not accept all that doctors tell us uncritically, for some doctors and hospital consultants do not assess conditions rationally, despite the emphasis on so-called 'evidence-based medicine'. Doctors must believe whatever the drug companies tell them about the medicines they prescribe and they do not have the time to research further into possible side effects, which pharmaceutical companies will minimize. Doctors are also not trained in nutrition and most have little or no knowledge of or respect for alternative therapies, so these facts need to be borne in mind when discussing health issues with a GP.

Patients need to be aware that medical diagnosis today is about finding the most appropriate synthetic drug to suppress whatever symptoms may be troubling them. Underlying causes of the disorder are rarely sought or investigated. There does seem to be confusion about compounds causing diseases, like the role of cholesterol in heart disease, because doctors are trained to look for one source for one disease affecting one organ. The human body does not work like that – it is a holistic organism. The training of doctors is based on the premise that treatment by synthetic pharmaceutical preparations is the proper and only effective way to treat any patient, other than by surgery. A patient presents with a particular symptom, or group of symptoms, which are taken to indicate disorder in a specific organ or system of the body requiring a particular drug to alleviate the symptoms: this is the current 'biomedical model of health', as it is called. Often there will be many drugs by different manufacturers all claiming to do the same job. If adverse drug reactions arise, there is usually another drug that can be prescribed to alleviate those. That is how most doctors and consultants operate. The idea that the body must be treated as a whole, and that mental or emotional issues impact dramatically on physical conditions rarely enters into the assessment of symptoms [356]. Doctors and consultants are so sure of their ground that they rarely take heed of what unqualified patients have to say about their own bodies.

Family doctors or general practitioners (GPs) are not held in the kind of esteem that they once were – even a few decades ago. Despite this situation, GPs in Britain pressured the government into establishing new contracts of employment that provided average salary increases of the order of 30%. With the track record of the medical profession, and more people taking themselves directly to hospital casualty departments or turning to complementary health practitioners, this is a shameful situation. Many patients are no longer content to take whatever medication their doctor prescribes without some information as to what it is for and what may be the possible side-effects. There is little point in enquiring whether naturopathic alternatives are available because the philosophy of the orthodox medical profession is that such compounds are ineffective and useless, so few doctors have any notion of their availability or action. As pharmaceutical drugs have become more powerful, they have also become more toxic, as the death statistics indicate.

It was only in the 20th century that herbal medicines were replaced increasingly by synthetic drugs. Before then, most non-surgical treatment was by herbal medication[10]. The drugs developed over the past century have saved innumerable lives of people who would otherwise have died, and reduced pain and suffering for countless others. These drugs have been so successful that there has been a tendency to use them rather more prolifically than is beneficial to patients in the long term. Most medicines simply suppress symptoms and relatively few cure or even treat the underlying causes of symptoms.

In general, it is more holistic and better for the patient to use natural or herbal remedies provided they are safe and effective because they are usually gentler in action and produce fewer side effects than synthetic drugs[11]. Pharmacological or allopathic drugs are really only extensively tested on humans by many years of application 'in the field' and seeing what benefits and what side-effects occur. The animal studies on which the 'safety' of drugs is based are of dubious value and additionally raise moral issues. For many medical conditions, naturopathic or herbal remedies are also available that may be equally effective, without the usual occurrence of adverse drug reactions. In a holist approach to life, this is clearly the way we should go, certainly as a first option.

In considering the relative merits of orthodox or alternative medicine we should not lose sight of the incentives given to doctors to prescribe certain medications. Much medical research undertaken in medical schools and hospitals is sponsored by drug companies so there are vested interests involved in securing favourable outcomes of research projects. Drug companies sponsor conference venues, they contribute to the funding of post-graduate students who, along with practicing doctors often receive financial help with travel expenses to attend conferences promoting their products, and they give small free gifts and free lunches to medical practitioners. They inundate practitioners with literature advertising the benefits of their products; such literature would be too unwieldy if it were to go into any significant detail about side-effects or interactions of the drugs with other prescription medication. This literature is usually the only reading a busy practitioner will have time to cover. The pharmaceutical industry, like every other international commercial enterprise, is just as much driven by the quest for profits (see Chapter 6). Ethical considerations are secondary and consumer satisfaction is of importance only in enhancing profits. A holistic outlook demands that we keep these facts in view and keep pressure on responsible bodies to see that complementary medicine is not ignored.

Government money spent on health promotion and disease prevention would be a far better investment than spending ever increasing amounts on treatment after the body systems go awry, particularly for schoolchildren. As it is, in Britain, health promotion departments have to spend valuable time and resources making applications several times annually for funding. Such proposals may be turned down, so the time and effort expended is wasted and a new proposal must be created. Health promotion has always been the 'Cinderella' of health activities – we need to put it at the forefront if we are serious about getting the nation healthy.

The banning of smoking in public places, for example, is a measure that should have been made law decades ago as soon as it was realized what harm was being caused by passive smoking. The Environmental Protection Agency (EPA) in America issued a 500-page report in December 1992 on Respiratory Health Effects of Passive Smoking that stated: 'the widespread exposure to environmental tobacco smoke in the United States presents a serious and substantial public health impact'. In February of 1993, one of the senior vice-presidents of tobacco company Philip Morris sent an internal memo stating: 'Our overriding objective is to discredit the EPA report . . . Concurrently, it is our objective to prevent states and cities, as well as businesses, from [implementing] passive-smoking bans' [263, p.31]. Here we have a clear example of the ruthlessness with which corporate business pursues its profit objective totally heedless of any damage done to people's welfare. In holistic living, these companies would be closed forthwith and their executives charged with corporate manslaughter. We shall hear more of capitalist exploitation in Chapter 6.

Until the 20th century, 'alternative medicine' was almost the only kind there was, and this has kept humankind alive, reasonably well and reproducing effectively for the past million years or so. Its effectiveness was validated by experience rather than by the parameters of scientific investigation. This is not to deny the huge advances in medicine that have been made though the use of science – but it is not the *only* way of deriving evidence. There are many disorders and diseases for which allopathic medicines are undoubtedly the most effective and perhaps the only treatment for a patient. There are times when surgery is desirable or essential. But a holistic naturopathic approach to medicine considers the state of the whole body in producing imbalance, and seeks to treat causes rather than symptoms. Alternative remedies are usually non-invasive (except for acupuncture), gentle, and have very few and usually no side-effects.

Perhaps the most damning indictment of orthodox medicine is the huge number of side-effects associated with allopathic drugs, frequently resulting in disability or death. There is a whole branch of medicine – iatrogenic medicine – that has been created to deal with just this subject of adverse drug reactions (ADRs). These have gained more publicity over the past decade partly because of a number of books on the subject [see, for example, refs 78; 248]. In a fairly recent British study, a group of pharmacists and pharmacologists found that 6.5% of hospital admissions in a sample group of nearly 20,000 patients who were sick enough to need to be admitted to hospital were due to adverse drug reactions. Of these, 0.15% died. This looks to be a very small proportion, but that represents some 28 patients killed in just two Merseyside (UK) hospitals over the six-month study period! If that were multiplied up over the whole of the UK for a year, and even more if we were to include statistics from America, it would reach a frighteningly high figure for deaths from drugs that we are being recommended to take instead of 'unproven' herbal and naturopathic treatments. These deaths are from approved orthodox medicines taken as prescribed. We should bear in mind that the pharmaceutical industry is made up of a number of the kinds of mega-corporations that have distorted our holistic world-view by capitalist economics (see Chapter 6).

Drugs are always tested on at least two species of animals before they are given to humans and the results are reviewed by the Committee on

Review of Medicines in Britain and by the Food and Drug Administration in the USA. Unfortunately for us, as pointed out earlier, many animals do not respond in the same way at all as humans to infectious diseases or to the drugs given to them to treat their symptoms [13, 14]. Indeed, Jonathan Balcombe and colleagues have shown that it is stressful to animals to even be in a laboratory where experimentation is carried out [17], and we all know how stress affects our body biochemistry. Because stress has such profound effects on all living creatures and plants, the results of any experiments carried out under these conditions are bound to be distorted and misleading, quite apart from the different species responses.

Even our genetically closest relatives, the non-human primates (NHPs), do not react in the same way as we do to pharmaceutical preparations. For example, hormone replacement therapy (HRT) was tested on NHPs before being given to women after the menopause. Animal studies indicated that an additional benefit would be that it would reduce the incidence of heart disease and stroke. However, it is now known that HRT slightly *increases* the risk of these diseases, and breast cancer. In fact, it is believed to have caused up to 20,000 cases of breast cancer over the past decade in Britain alone [135]. Even such a widely used drug as aspirin may cause gastrointestinal bleeding in humans and for this reason is frequently replaced by the potentially highly toxic compound warfarin to thin the blood; but aspirin, although it may cause gastric bleeding in humans, is actually teratogenic (causes birth defects) in monkeys. In a study from St. Mary's Hospital Medical School in London it was found that among rodents and primates, zoologically closely related species exhibited markedly different patterns of metabolism [40].

A recurring theme of many complementary and orthodox health practitioners is that a climate of fear and the resultant stress it produces in individuals, together with unhealthy lifestyle choices in food and drugs, have largely contributed to many of the ailments and diseases that have become prevalent in the 20th century. The importance of stress or fear as a provocative agent of illness has been attested to by many psychologists and spiritual healers as well as by orthodox clinicians. Many different kinds of individual life circumstances can produce mental anxiety or stress that leads to physical symptoms of ill-health: fear of ill-health, of hospitalization or surgery, of becoming disabled in old age; fear of having to look after an aged dependent; fear of redundancy and not having enough financial resources for food or clothes, or not being able to pay fuel bills or make mortgage payments; fear of an abusive drunken spouse or debt incurred by a spendthrift partner's gambling habits. Such fear or anxiety or stress is now recognized by the medical profession as a significant contributor to loss of time at work and ultimately to major organic diseases like heart disease and cancer. If people have a low feeling of self-worth, or are simply bored, they will tend to over-eat or drink alcohol and that will cause physical ailments of their own. Obesity increases considerably the risk of death by stroke or heart disease, and often leads to diabetes with resultant blindness or limb amputation. But overeating is frequently the result of attempted compensation for stress. We shall meet the subject of fear again in the next chapter in respect of the fear of self-reliance suggested by Erich Fromm's *Fear of Freedom* [110].

Children's health and welfare

Parenthood carries with it an enormous responsibility to see that life is better for one's children than for oneself. Parents are a child's first teachers, and constant parental input increases a child's security and happiness, develops thinking and communication skills, and diminishes fear and apprehension [240]. If women feel that they cannot stand being in the house with a baby all day (as I have heard many say!), then why do they have children? A child is not a toy to be played with when we feel like it but given over to child-minders or the state to look after when their needs become inconvenient. Raising a child demands making a commitment for at least a couple of decades to put the child's welfare first.

That bond between a baby and its parents that is built in the first few years of life is absolutely irreplaceable. That is not to say that if misfortune – a broken marriage, death of a parent or financial constraints – forces parents to be out at work each day, a child is necessarily irreparably damaged. But the chances that the child will grow into a happy, contented and fulfilled adult are lessened [240]. We need look no further than our society today to see a correlation, at least statistically, between the number of broken homes and overworked parents and the decline in behaviour and performance of children at school. Of course, this is only one factor and, as we shall see, things like diet, sleep and play also contribute to children's behaviour. The commercial demands of life today mean that parents have less time or energy to devote to child-rearing, especially if both parents choose to or are forced to work in order to survive [175].

For a similar reason, it is undesirable that homosexual couples be allowed to adopt children – not because of any moral or religious criticism of their relationship but because children optimally need input from male and female parents to get a balanced, and what is regarded as normal, view of the world. The argument is a purely rational one as to what is in *the best interests of the child* whatever sacrifice the couple may have to make for the sake of their relationship. The same principle applies to IVF treatment for single women or those in their later years of life. There is too much emphasis on the rights of women to have children and too little emphasis, within a holistic world-view, on the duty of care in the upbringing of those children. The relationship between rights and duties is discussed further in Chapter 5.

Even before children are born it is particularly important for the welfare of the foetus that expectant mothers refrain completely from smoking, drinking alcohol or drug taking. Even prescription drugs should be kept to a minimum, especially in the first three months of pregnancy. It is now well established that passive smoking of other peoples' cigarette smoke is harmful to all, especially in a confined area like the family home; so if parents care about the welfare of their children as well as themselves they will not smoke in their presence.

Children need to be guided into those activities that will give them health and pleasure but will avoid pain in the immediate or long-term future, both for themselves and for those within their circle of future acquaintances. Children also need to have the courage to resist peer pressure to join in activities they know will do them harm. If parents smoke and drink alcohol – certainly to excess – they are in no position of moral authority to condemn their children for experimenting with so-called recreational, but in fact deadly, drugs like cocaine and ecstasy. The philosophy of moderation in all things is one that surely needs to be inculcated into our schoolchildren at as early an age as possible, but necessarily following not only instruction but also by example. The holistic message needs to originate from home and be supported at school. For the individual following a path towards holism, 'health' means catering for the body, mind and spirit of self and of others.

Anyone who has been in teaching over the past few decades will tell you that maintaining discipline in the classroom has become increasingly challenging. The accent on 'children's rights', the removal of effective sanctions against those who misbehave, and the introduction of mixed-ability teaching have not helped this situation at all. There are many factors contributing to lack of attentiveness in the classroom. Some of these will be covered in Chapter 8 on Education, but the focus here is on health issues. Putting disabled children in classes with those who are not impaired does a huge disservice to all. Disabled children need extra attention and in classes of thirty or more pupils there simply isn't the time. They are also frequently ridiculed by the other able-bodied children, and every day they may be made aware of their academic shortcomings because of their impairment.

Some 12% of American children now suffer from ADHD (attention deficit and hyperactivity disorder) and rates of occurrence are increasing throughout the developed world. Nearly the same proportion suffer from the related group of disorders dyslexia (difficulty in reading), dysgraphia (difficulty in writing) and dyspraxia (difficulty with physical coordination); and an increasing proportion are also exhibiting autism (difficulty in communication, coordination and focused attention). It is still a controversial issue as to whether or not some cases of autism tragically arise as a result of immunization with multiple vaccines.

Dr Richard Mackarness was one of the first to draw attention to the almost instantaneous adverse effects of some foodstuffs on both body and mind in children [233, 234]. Support groups were set up and were much needed because general practitioners at the time were very dismissive that any such problem could exist. Doctors are still loath to attribute any illness to vitamin or mineral deficiency, or to exposure to substances other than known common allergens. A rash of self-help books also appeared in the 1980s to give advice to anxious parents [37; 264; 307][3]. Strictly, the term 'allergy' refers to an exaggerated response of the immune system to an invasive protein, but hypersensitivity responses can be provoked to almost anything in some individuals[12].

Organic food tends to produce fewer effects. It is quite alarming that in 2004, the American Academy of Pediatrics reported on their website that '1 in 6 children are diagnosed with a developmental disorder and/or a behavioural problem' [285, p.4]. The cause seems to be a combination of a malfunction in genetic programming (so easily damaged by toxic chemicals) coupled with unholistic $20^{th}/21^{st}$ century living, both nature and nurture playing their part. Bruce Lipton and others have shown [221, 222] that while the structure of DNA is determined at conception, its function can be modified by environmental factors.

It is certainly true that, to be healthy, children need far more exercise outdoors in green play areas than they are getting today. In Britain, too many green spaces to play sports have been converted into supermarket car parks and it has become much more hazardous for children to walk about freely on their own, partly from density of traffic, partly from threat of abduction. It is easier for children, and parents consider it safer, for them to stay at home and watch television or play computer games. As a result we are creating a generation of overweight and unhealthy children and we see the decline of what Sue Palmer calls the 'free-range child' who once used to roam free and safe and happy [285, p.59]. Other authors have also commented on the adverse effects on children of keeping them indoors too much of the time [228].

These were the woods the river and sea
Where a boy
In the listening
Summertime of the dead whispered the truth of his joy
To the trees and the stones and the fish in the tide.

From Poem in October by Dylan Thomas (1914-1953)
[Collected Poems, 1934-1952, Dent, London, 1952]

Walking or cycling to school has now largely been replaced by parental chauffeuring or 'bussing', partly from concerns for children's safety. But the traffic congestion and environmental pollution caused by all this commuting is one of the reasons given as to why the walking and cycling is no longer safe! The other reason given for bringing children long distances into bigger schools is that they can provide the fullest range of career choices to as many pupils as possible. This is probably a misguided concept. Children who want to go on to further education can wait till age 15 or 16 to take advantage of the full range of career opportunities offered by a college. What they need as children is a secure, friendly environment where they can acquire a sound education in the basics of reading and writing fluently in their native language and the elements of calculation (so often sadly lacking today) and simply a glimpse of what possible career paths would be open to them from taking different subjects. Age 13 or 14 is far too young for a child to be making career choices that are essentially irreversible. If children are not getting exercise traveling back and forth to school it is even more important that physical education or games classes not be sacrificed to cram in more academic study in order to meet examination pass targets.

Getting enough sleep is particularly important for children who need 8-9 hours of sleep a night during their teenage years and 10 hours a night in primary school. The attractions of television and computers and the pressures of schoolwork are driving children into workaholic and playa-holic automatons, and many adults are getting caught up in these pressures too. Some parents are too ambitious for their children leading to 'hyper-parenting' – the compulsive drive to push one's children as much as possible in as many areas as possible. It is even more important for developing children than for adults to keep the components of healthy living in balance [30]. Honoré has commented on the need for children to avoid this sort of frantic pressure to which their parents, more often than not, will be exposed in their professional lives [175, Chap 10]. As he says, it is better to do fewer things but have the time to get the maximum fulfilment out of each rather than rushing through to get to the next activity: 'be sure to cultivate the art of doing nothing'. Of course, some students will gleefully take this to extreme and spend their time carousing, but most students are serious and those who overdo activities now probably outnumber the 'perpetual students' who eke out a degree course for many more years than necessary. Holistic living is about the fulfilment of wholeness: we lose so much by rushing from one experience to another without giving ourselves time to immerse our selves in each. Only then can we enjoy the full benefit of each place we visit in our journey through the landscape of life.

Educational psychologists tell us that the lack of concentration and propensity for indiscipline shown by schoolchildren is in part due to insufficient sleep, often due to hyperactivity, and that inadequate nutrition as a result of eating too much junk food, too little exercise in the open air and insufficient meaningful human interaction (with parents or other family especially) are other major contributing factors. Oxford physiol-ogist Dr Alex Richardson regards the junk food to which so many children are now addicted as essentially toxic [313]. Thanks to concerted targeting by the marketing industry, children have been encouraged to eat low quality processed foods and drinks that provide instant gratification instead of foods that may seem less appealing at the time but which build young bodies sturdily for the future. 'Treats' of sweets and fizzy drinks full of additives do children no favours at all. Some leadership from government is needed to curtail the activities of the advertising industry in their efforts to indoctrinate children with the desirability of these harmful products.

Children have been indoctrinated and tempted with junk food for decades and reversing this trend is not going to happen overnight. A review of studies of children's nutrition carried out by the British Nutrition Foundation in 2005 found that 50% of children had a marginal intake of vitamin A and 75% had a marginal intake of zinc, both of which are essential nutrients [285, p.27]. If these nutrients are deficient it is unlikely that the intake of many others will be adequate. Sensible, nutritious eating habits must start in infancy. Surveys indicate that 20% of school-children in the UK still eat no fruit or vegetables and are overweight if not obese. Encouragement to eat healthily has to start in infancy so that children are used to the different tastes and flavours and do not baulk when presented with broccoli at the dinner table. Recent research also suggests that lack of sleep is one of the reasons for obesity in children. In a child who has insufficient sleep the hormones are turned on to the stress state and encourage eating, especially of energy-giving, sugary foods providing calories that are not burned off by exercise.

If we want to build a healthy society in the next generation we need to start addressing some of the many problems in the raising of children now and see that they have a healthy holistic lifestyle! As parents and teachers it is within our power to influence childhood development in the right direction, provided we have the relevant information. But we also need availability of nutritious food and safe green play areas, some government restriction on brain-washing by the advertising industry encouraging children to eat junk food, higher penalties for selling drugs (including tobacco and alcohol) to children and a societal commitment to see that we start moving these factors in the right direction.

The bottom line from these facts about our increasingly hedonistic, materialistic and synthetic society is that we all have a responsibility as individuals to take care of the health of our bodies for our own sake as well as for the sake of those relatives or friends or health service personnel who will have to look after us if we become ill though our own selfish indulgences. Most of all, it is incumbent upon all of us as parents to accept responsibility for the care of the children we bring into this world and to see that they are loved and nurtured and guided to the very best of our ability and not abdicate this responsibility to schools or the state.

The mind

Stimulation of the mind goes hand-in-hand with healthy function of the rest of the body. If we try to achieve a contented state of mind and focus on the beauty that surrounds us, this exerts a positive effect on the whole physical body and predisposes us to good health. There are numerous scientific studies of the beneficial effects of *placebos* – medicines with no physiological effect in themselves but producing positive effects because those taking them thought they were being medicated and effectively were willing themselves better. Optimistic, happy people are healthier than those who are negative and pessimistic [240]. If you believe that something will make you ill, even if it is physiologically harmless, it can produce a negative effect on the body – such effects are known as *nocebos*. Positive (or negative) energies generated by the mind communicate this to the cells of our body and predispose them to healthier (or disordered) function [222]. This is what Adrian Harris calls a Somatic Philosophy – an awareness that mind and environment impact continually on every cell of the body [157, p.152] as indicated by the researches of biologist Bruce Lipton [222]: body, mind, spirit, environment are one.

Harris believes that we will not achieve a sacred ecology until we come to believe in this principle and let it infuse into everything that we do.

The positive effects of prayer and an optimistic outlook and, conversely, the damaging effects on physical health of negative emotions are well documented. This does not mean that positive thought alone will produce a cure from every disorder, as Christian Scientists believe. Again, empirical evidence indicates that this is certainly not the case. But what is equally clear from practical experience is that self-belief and being surrounded by loving care from relatives, nurses and other carers *is* a major contributory factor to recovery from illness [331]. Developing serenity within the self helps to induce this in others [144]. As Bruce Lipton says: ‘Our positive and negative beliefs not only impact our health, but also every aspect of our life’ [222, p.143].

The term ‘placebo’ is sometimes used disparagingly by rationalists to dismiss healing claims by complementary therapists. Those with a restricted, materialist world-view ascribe any beneficial effects of such treatments to the active imagination and strong will of the recipient. But spiritual healing and other complementary or alternative medical techniques are not placebos in this negative sense. While the healing is ultimately self-generated within the patient, an effective spiritual healer will be able to summon the universal spiritual energy or *chi* to stimulate the patient’s own curative function. The energy is channeled through the endocrine and other glands of the patient. Healers believe that each gland is associated with an energy vortex or *chakra*[13]. Rationalists who dismiss such notions would do well to remember that acupuncturists produce anaesthesia by using meridians of energy in the body that have been known to eastern practitioners for thousands of years, but have no equivalent in western medicine.

A holistic outlook means working towards real, deep-down joy or satisfaction, not the quick fix of a cigarette, a glass of beer or a shot of cocaine that only makes us want more of the same, without satisfying the adrenaline rush for more than a fleeting interval. If we were wholly rational beings without emotional impulses, thinking of the consequences should deter us from searching for many of these new, seemingly ever more exciting highs. When we look back, we realize these were no more than transitory, superficial pleasures that took our lives no further forward but, rather, a few steps back or, perhaps, into the abyss, dragging others down with us. So many of our actions arise from impulses, the consequences of which emerge only later with the wisdom of hindsight. Although the physical parameters for health are much the same for all people regarding diet and nutrition, what constitutes pleasure will be enormously variable from one person to another. All that can be said is that we must seek out those things that give us pleasure, provided others are not inconvenienced or injured, and avoid those things we know make us unhappy overall, even in the long term[14].

Many sporting activities, although primarily physical, also demand much of the mind; a dejected state of mind can often be lifted by physical activity, especially if shared with others. For those who are unable to participate in physical activities, pursuing a hobby, learning a language, joining a study group such as those run by the Workers Educational Association (WEA), the University of the Third Age (U3A) or the Learning in Retirement Movement (LIR) in America will give social contact as well as a chance to learn new skills, not necessarily for any career advancement but simply to keep the mind active. Organizations such as the U3A and LIR have been created so that the increasingly aged segment of society can keep their minds (and bodies too if they are able) active in the later years of retirement. Human individuals have an innate need to belong, to be part of something greater than themselves: this is why we join sports clubs, or religious or political groups, or go to evening classes, as well as for the mental or physical challenges. In such groups we can find stimulation for the body, mind and soul. The restriction in funding during the 1980s in Britain of further education for adults took away from people the opportunity of fostering community spirit as well as depriving them of a social outlet.

We cannot live a life of unbroken bliss in the real world, and some disappointments, frustrations and sad events are inevitable, but if our outlook is positive we are better able to regard such events as learning experiences. It is particularly important to look on the later years of life as a time for reflection and to catch up on as many as possible of the things we did not have time or money to do when we were young. There is no reason why we should expect our later years to necessarily be years of inactivity and ill-health. At any age, a positive outlook is always productive and uplifting and continuing mental activity delays the onset of the cerebral aging process.

The newly developed medical discipline of psychoneuroimmunology (PNI) recognizes the interrelation between the nervous and immune systems and how stressed states of mind produce physical illness. Conversely, relaxation, contentment and fulfillment predispose towards wholeness and health and make essential contributions to happiness. An individual's mental and emotional states have a huge impact on both the cause and the cure of illness or physical imbalance within the body. The quest for material wealth and power may seem tempting goals, but many of those who have one or other, or both, often find that any happiness they bring is short-lived. Lottery winners frequently find that after the initial euphoria is over, money does not in fact buy happiness. Though people may have a genetic predisposition to a jolly or melancholic personality, happiness is something that can be cultivated by all. As we shall see in Chapter 3, the focus of eastern mystical philosophy is the search for happiness by finding ways to quell discontent and striving for often unattainable goals.

We all know how much more fatigued we feel after a boring day at the office than if we have been stimulated through being constantly busy. Of course, there is busy-ness and there is the often frantic activity that seems to be demanded in so many occupations today. This situation contributes increasingly to diseases associated with stress. We need times of rest and relaxation with an engrossing novel, or reading poetry, listening to music, or getting out of the city clamour into unspoilt countryside so that, with practice, we are transported to another dimension. Playing a sport, or finding a quiet place at home to become lost in our own thoughts, pursue a hobby, or to share time with friends or family – these are essential components of holistic living for which we need to make time in our pressured 24/7 lives that business and the ad-men would have us lead. However much energy we may feel we have when we are young, making time for whatever we find enjoyable and relaxing is still an essential component of a holistic lifestyle. The time spent at work is then likely to be more productive, even if it is shorter.

In our younger days we are likely to have a family around us, but again, in the 21st century urban setting, families are dispersed by work pressures and broken relationships. Commuting adds its own pressures to those of demands for ever-increasing profits in business: time spent in commuting is family time squandered quite apart from its damaging effect on the environment. Over the past decades these commercial pressures have led to breakdown in very many family groups so that a child being raised by its natural mother and father with the extended family of grandparents and other relatives nearby is becoming an increasingly rare phenomenon. Women may want to work away from the home for their own satisfaction, or may be forced to do so by financial pressures. Men are the ones whose job often demands working away from home for long periods. All of this weakens the nuclear family unit as parents spend less time with their children, who are left to amuse themselves by watching television and playing computer games. We should not be surprised if basic standards of literacy and numeracy are found to be declining, whatever government statistics on examination results may suggest to the contrary. The mental stimulation provided by computer games is far inferior to that provided by reading because the latter requires image-making by the child. Mind issues will be explored further in Chapter 2.

Many people have made the conscious decision to 'downshift' to less stressful living with fewer material possessions but providing a greater quality of life by focusing on the simpler pleasures. This is a radical change of direction for many but, having experienced the contentment that this lifestyle brings, they have no wish to return to their previous existence. A consequence of this is often a greater appreciation of the spiritual values of life.

The spirit

There are many for whom material needs have been satisfied and recreational wants catered for: now their search is for something deeper and more meaningful that will impact on their lives in real rather than just abstract ways. The sense of wonder and awe at the depth and breadth of human creativity, and even more at the grandeur of the natural world, is at the core of our humanity. Appreciation of the aesthetic dimension of being is one way that many find fulfilling as a means of escape from their material surroundings and at the same time it can provide contentment and inner peace. The only purpose of beauty in the world is surely to provide happiness and uplift for humankind. It serves no practical purpose, yet no society could call itself civilized without it. Indeed, its very existence is only within the human mind and soul. Some find their beauty within the natural world:

> Beauty being the best of all we know
> Sums up the unsearchable and secret aims
> Of Nature
> Robert Bridges, *The Growth of Love*, 8 (1876).

Others find their spiritual joy in poetry, music, or works of art, sculpture or architecture. Beauty, as we might expect, has been an inspiration for poets, painters and composers for as long as there have been artisans, but in the Romantic period of artistic creativity in the 19th century emotional expression became particularly important.

A thing of beauty is a joy forever:
Its loveliness increases; it will never
Pass into nothingness
John Keats, *Endymion*, 1.1 (1819)

Keats' vision here, that the concept of beauty is held forever in the universal spiritual realm, is an echo of the Platonic Ideas. That it increases in intensity every time it is experienced as well as when it is created, accords with the view of spiritualists. Our concept of beauty in the material world around us is a reflection of the eternal and infinite Beauty held within the spiritual plane. All of our thoughts and actions, both positive and negative, add to the store in the spiritual realm. This is why, with a holistic vision, we must reinforce positive thoughts and actions, and generate beauty at every opportunity in our daily lives.

The realm of music provides perhaps the most intense and universal source of spiritual joy, and may well improve cognitive skill too [266]. For both the composition and the appreciation of music involve the emotions and the intellect of composer and listener. As conductor and pianist Daniel Barenboim commented in his first Reith Lecture: 'Music can and should become something that is used not only to escape from the world but rather to understand it' [18]. Music is a medium that transmits thoughts from the mind and feelings of the soul from composer to listener. Music is unique in that it usually presents no specific picture to the listener, unlike painting or literature, and perhaps this 'creative' aspect demanded of the listener makes listening to classical music the most intensely spiritual of aesthetic experiences. Much music is described as 'absolute' – that which is not intended to convey a particular picture. During the Romantic period, much music was composed that did intend to portray a scene in sound. However, it may be said that the meaning of pieces of music is generally implicit while that of works of art or poetry is, in many cases, more explicit. Where meaning is explicit in the setting of vocal music, the music enhances the moving and thought-provoking experience of the text. Popular, contemporary, jazz and folk music can make us feel understood, that we are not alone and can inspire us to change the way we look at the world. One of music's greatest strengths is that it does not present the potential language barrier of literature.

Some 19th century dramas featured incidental music such as we find almost invariably today in films and often television productions. As we might expect, perhaps the greatest of all dramatists, William Shakespeare, has inspired many pieces of music from his plays, as has Cervantes' hero Don Quixote. Incidental music scores written for films have become popular today as stand-alone works and they now represent a musical genre all of their own. The presence of music in providing the atmosphere of a scene in movies is quite indispensable, so much so that in the days of silent films, pianists were engaged to reflect and enhance the drama of the action on screen by their piano playing. As we might expect, sacred texts too have inspired many moving musical settings. As well as inspiring much poetry and painting, the beauty of the natural world has engendered the composition of numerous musical works.

Music of all types represents a spiritual communion between the souls of composer and listener and it is generally left to hearers to create their own vision. This is why so much modern music (and art and poetry) is more difficult to appreciate: it is art for the artisan, creation for the cognoscente. It is much more difficult to derive spiritual uplift from much modern art, music and poetry than from the poetry of Keats or Wordsworth, the painting of Constable or even Turner, which is quite impressionistic, or the music of Mozart or Vivaldi. The humblest listener, reader or viewer can derive pleasure from these earlier art-works because a degree of understanding is readily achievable; those who want to find deeper meanings will find them there too. Of course, art must progress and find something new to say, like every other aspect of life. But newer expressions of artistic feeling, if they are to serve any purpose for the general public, must remain accessible. The expressions of confusion, bewilderment, dismay, even disgust at some contemporary artistic offerings provide clear evidence of how they are generally perceived.

Apart from its implicit transmission of feelings, soul to soul, music is unique in another way too. Poetry and painting may represent almost any subject at all – except one another – but they cannot effectively represent music. Poetry and painting are essentially visual pleasures, while music is aural. Music differs also in that it can and does often represent poetry and painting. There are numerous examples that can be enjoyed even without any deep understanding of either musical theory or even the texts of poems, though translations do make the musical works more meaningful. Paintings too have been the direct inspiration of a number of musical works. These art-works represent an expression of spiritual holism, where artistic inspiration within one medium is renewed and enhanced by expression in another.

Many creative artists, composers and writers, including the present author, believe that the creative process, though expressed by the individual, has its source in the spiritual domain. The Austrian composer Gustav Mahler (1860-1911) was one who saw the process of composition as part of this mystical interaction. When speaking of his Second Symphony, which is popularly known as *The Resurrection*, he said: 'Creative activity and the genesis of a work are mystical from start to finish, since one acts unconsciously, as if prompted from outside, and then one can hardly conceive how the result has come into being' [29, p. 196] and 'For me, the conception of the work never involved the laying down of a process, but at the most of a feeling ...The parallelism between life and music may be deeper and wider than we are yet in a position to understand' [*ibid.*, p. 204]. In an interview on U.S. TV, the Russian born composer Igor Stravinsky said of his composition of *The Rite of Spring*: 'I heard, and I wrote what I heard. I was a vessel through which *Le Sacre* [*de Printemps*] passed'. These composers believe that such creative inspiration derives from an external spiritual source and that they tap into that spiritual domain in their compositions. Cellist Steven Isserlis similarly sees a divine origin for inspired composition. In an interview with Oliver

Condy for the BBC's *Music Magazine*, Isserlis was asked about his preparation for performance as to whether or not he listened to other recorded performances. Indicating his preference to go back to the manuscripts themselves Isserlis commented: 'Why get your instruction from a vicar when you have a chance to talk to God?' [187].

Music has been used since earliest times as an accompaniment to religious ceremonies and gatherings for rites of passage, and music is often an integral part of tribal and folk traditions[15]. Such ceremonies frequently involve dance as well, as another expression of spiritual communion. The spiritual connection is true not only in the creative arts. When we make some inspired leap of mind to the solution of a mathematical or scientific problem, we often scarcely know from where it came. All of our training seems inadequate to resolve the problem, until the light suddenly shines. Looking over what we have written, we may not be able to trace a logical path: the ideas have simply materialized from air, as it were, channeled through mind and body, just as Mahler expressed above. In other spheres, we can touch the soul of the universe through our sense of awe and wonder at the beauty and grandeur of Nature, and through the fulfilment of human loving. It is not necessary to have visions of the divine in order to become aware of a connection with our spirituality. It is not the dogma and ritual of organized religion that provide the spiritual dimension of holistic living but this experience and expression of innate spirituality. This should be the true message of any religion – to search for inner spirituality in harmony with that of the universe.

Many healers use music as an accompaniment to their healing; dentists use it to soothe their patients, and there are surgeons who claim that it is beneficial to the smooth running of operations in the theatre; it is especially good for patients if surgery is to be performed under local anaesthetic or acupuncture. For the rest of us, once the ear is attuned to music, it can be the most powerful agent for reducing stress and producing relaxation and joy.

Immersion in the aesthetic world of music, poetry, painting, or natural beauty allows us, for a while, to become independent of the physical world that rules our everyday lives, to seek and hopefully find joy in a personal psychic dimension of our own[16]. We can do as the mystics of Eastern religious philosophy encourage us to do and lose ourselves in our own meditative paradise within the material world:

> We sit together,
> the mountain and I,
> until only the mountain remains.
> Li T'ai Po (701-762)

Summary

Basically it is incumbent on each one of us to look after our health, as far as it is within our power to do so. Eating nutritious food, taking moderate exercise, avoiding all drugs if possible, following a stimulating and worthwhile job or profession, getting an adequate amount of sleep and not allowing ourselves to be pressured by employers or advertising media to try to live over-active lives, cultivating stimulating friends who share our world-view would all enable us to live happy and holistic lives. If we do not help ourselves and exercise moderation and self-control, we cannot expect others – and particularly the state – to come to our rescue and bail us out of trouble when things go wrong. We cannot live hedonistic, irresponsible lives of excess in any aspect and not be mindful of the later consequences. The whole of life is a learning process dedicated to the achievement of happiness. As we have seen, that involves avoidance of (later) pain as well as achievement of (immediate) pleasure. One of the most holistic ways of achieving the third component of happiness is to do whatever we can to provide happiness for others.

CHAPTER 2

MIND

The nature of the human mind is the unique quality that distinguishes us from other animals. We are the only animals that are capable of speech or of rational thought or of moral behaviour that goes deeper than genetically programmed altruism that benefits the perpetuation of the species. Indeed, Pradervand [302] and many other mystics and pagans see consciousness as the all-pervading fundamental process of the universe. It is therefore worth exploring the insights into the workings of the human mind that some of its principal investigators have achieved over the past two centuries. Some understanding of the way the mind works gives us directions our paths might take towards a holistic lifestyle.

The subject area

The mind became a subject of study in its own right as recently as the 19th century, though philosophers had long conjectured about the nature of knowledge, and how human thought processes differed from those of other animals. The study of the mind as a separate discipline began in the mid-19th century initially out of a fusion of physiology and philosophy[1]. A key textbook in the early years of the subject was *The Principles of Psychology* (1890) written by American philosopher and psychologist William James (1842-1910), brother of novelist Henry James. The study of mind and soul in philosophy, psychology and religion were all closely interwoven still in the 19th century. Another of William James' works, *The Varieties of Religious Experience*, published in 1902, is still a classic in its field.

In the first century of its existence, psychology split into various subdisciplines[2]. The Russian scientist Ivan Pavlov, followed the example of the early German pharmacologists and used animal models for his studies, opening up the field of *animal psychology*. Pavlov won the Nobel Prize in 1904 for his research on the digestive system. The results Pavlov obtained with his dogs established the fact that we will respond in certain predictable ways when given the right stimuli after conditioning. After salivating when they were fed regularly, accompanied by the ringing of a bell, Pavlov's dogs began salivating just when the bell was rung. The whole of the advertising industry is based on provoking a similar conditioned response from consumers. Humans are genetically programmed to seek pleasure and self-esteem. Once we start acquiring material possessions we know from our previous life experiences that new material goods will give us pleasure and make us feel that we are on a par with our peers, or even superior to them, if only for a while. In Pavlovian fashion, the advertising industry manipulates us with the prospect of more and more material goods that will make us go on feeling better and better, or so they imply. They even use psychologists to help them present their commercials most effectively.

In the 20th century, interest developed in how the mind changed as we grew up, so the fields of developmental, educational and child psychology were established. The advertising industry uses such professionals to manipulate children's buying habits. This has become of huge importance with the increasing problems observed in the behaviour of children, as pointed out by Sue Palmer [285]. Although the title of her book, *Toxic Childhood*, does not seem encouraging, in fact it is full of suggestions as to how we could change prospects for the present and future generations. Children are indoctrinated with promotions for junk food and drinks that they should have because they are pleasant to eat and drink and because their friends have the same in their lunch boxes. The food is dressed up in colourful packages, and favourite film, cartoon or Disney characters are used in the promotion to create a friendly image. As Sue Palmer says, 'for many Americans the mere sight of a can of Coca-Cola excites activity in sections of the brain associated with feelings of self-image, memory and cultural identity' [285, p.24]. Advertising executives exploit the pester power of children to persuade their parents to buy toys or clothes, if only to keep up with their classmates. This kind of brainwashing and indoctrination is something we need to be on our guard against in holistic living because we are being manipulated and it is effectively removing our rational choices – like Pavlov's dogs, we are being programmed to respond in certain ways and buy goods that we are made to feel we want. This is a negative application of Pavlov's work.

However, we can also take a positive view of Pavlov's results. If we cultivate optimistic, happy and spiritual friends, if we go to places that give us pleasant experiences, eat food that will not make us feel ill later, refrain from drugs like alcohol or tobacco that may give us transitory pleasure but ruin our health, if we make the effort to seek out the kinds of aesthetic experiences that we enjoy, then we surround ourselves with a positive and uplifting aura, and the prospect of further experiences of the same kind reinforces our buoyant outlook whenever we encounter these situations again. This is especially true of the experiences of childhood.

Where psychology has, for the last century, followed the lead of psychoanalysis and focused on the dysfunctional mind in depression or anxiety, there is an increasing interest now in how we can achieve happiness and satisfaction in life: this has created a new field designated as *positive psychology*. Many experimental studies have shown that happy people are healthier and more successful, and each of these life components feeds off the others: health, happiness and success go together. Conversely, anxiety, stress, ill-health and sadness (or, in extreme cases, depression) also go together. Happiness needs pleasurable experiences, like good food, stimulating friends, exciting sex or enchanting music. It also needs the absence of pain or anxiety; and a third component, satisfaction or fulfillment – the achievement of goals. All three of these comprise happiness to the psychologist: pleasure and pain are about how you feel at the moment (an emotional experience) while satisfaction is a reflection of how you see your life at present in relation to the goals you set yourself (a mental or rational assessment). According to Paul Martin, the most important component of the three in contributing to happiness is the strength of social relationships, the foundations for which are laid in childhood [240].

It is in the third of these that the holistic attitude becomes most relevant. Completing an arduous task or living an ascetic life may not give a great deal of pleasure but may well provide much satisfaction. Eating chocolate or taking drugs may give immediate pleasure, but this will soon wear off and the long-term satisfaction will be minimal; if it makes you fat or ill, it will actually cause you pain or distress. This is a vitally important principle to communicate to children – what the psychologists call *deferred* or *delayed gratification*. It's the philosophy of the Greek Epicureans – to be selective in their pleasures – as opposed to the Hedonists, who took whatever immediate pleasure they could, irrespective of long-term adverse repercussions. Not everything will provide pleasure (or pain!) immediately; sometimes the effects take longer to appear, as with the ultimate rewards of schoolwork in childhood, or the devastating illnesses caused by smoking, drinking alcohol or other drug-taking. Once the lesson has been learned, true happiness provides greater motivation than money or the superficial pleasure of instant gratification with unpleasant long-term consequences. But as Paul Martin says: 'Children can all too easily develop a lifelong habit of relying on short-term pleasures rather than learning to attain satisfaction' [240, p.18]. As we shall see shortly, other psychologists too have stressed the importance of social interaction as the defining event in shaping personality and I shall say more about that in Chapter 5.

As discussed above, our materialist capitalist society and its advertising minions are sustained by incessantly brainwashing us into believing that material possessions will make us happier and add to our social status. In fact, they may provide pleasure and instantaneous gratification, but this is usually short lived. Once our needs are satisfied – for food, clothes, shelter, human companionship and some material comforts – more of the same may help our image amongst our peers, but it does little else to provide happiness. 'Retail therapy' is a fleeting pleasure. It is the satisfaction of goals that provides the most meaningful contribution to our happiness. One of the hardest lessons that children have to learn is that of delayed gratification. Doing something now, like studying for exams, that may entail sacrifices will enable them to get better jobs and support themselves and their families, and thereby make them feel better later. The whole of education is like this; so while it is important to make the experience as enjoyable as possible in the present, children need to be aware that learning requires effort that will be rewarded in the longer term. Children need to appreciate that it is not the outer world that provides our contentment but how we perceive it and relate to it. It is the transcendence of materialism (though not its complete rejection) that leads to happiness, which is the fundamental principle of a holistic spiritual philosophy.

The importance of the unconscious

For more than a century now, psychologists have accepted that the unconscious mind plays a vital role in shaping human behaviour. The first psychologist to describe this specifically was Sigmund Freud[3], giving more precise form by Freud to an idea that had been referred to even in Greek drama. Initially Freud divided the mind into three functions: the *conscious mind* that is aware of its thoughts and actions; and, underlying this, is the *unconscious*, which holds those memories and feelings we never bring into consciousness, because they are too painful, or they are socially forbidden (such as the Oedipus complex, where young boys have incestuous feelings for their mothers but develop a love-hate relationship with their fathers). This aspect of mind represents what is often called the 'darker side' of the personality. Such thoughts are hidden from access by a process Freud called *repression*. The third division of mind is the *preconscious* (popularly called the *subconscious*), which acts as caretaker for all the information, feelings and memories that we can call on and bring into consciousness at any time but which are not in immediate mental focus. Freud agreed with the majority of psychologists and philosophers in regarding the attainment of pleasure (the 'pleasure principle' he called it) as the purpose of life [109, p.16; 199, p.320; 240]. Following the exploration by John Locke (1632-1704) of the nature of knowledge and the purpose of human existence [226, XXI, 41], this idea of the importance of happiness became enshrined in the American Constitution as a fundamental right of all men: 'the right to life, liberty and the pursuit of happiness'.

Freud later envisaged a different three-fold division of the mind or personality as the *id*, the primitive unconscious drives we are born with; the *ego*, which is essentially the 'self' participating in the world; and the *superego*, a kind of moral monitor or conscience controlling the impulses of the id.

The key principles to emerge from Freud's studies (of mainly neurotic middle-aged, often Jewish, Viennese women) were that behaviour is determined to a large degree by the unconscious mind of the individual, and that the motivation is very largely sexual from childhood onwards. This basic, essentially sexual, psychic energy that drives our actions Freud called the *libido*. Freud found that people would often betray in their word association or speech something that was hidden in their unconscious mind ('Freudian slips' or parapraxes). This is a feature of the human personality that we all make use of in our social interactions along with unconscious non-verbal communications like body language.

If patients thought (even unconsciously) that some event would happen, it often did – what we describe as 'wish fulfillment', though the event may have been more feared than wished for. There is a school of thought that maintains that the unconscious or conscious mind is a powerful force in influencing events in the physical world. Fears of illness can produce the very disability that was imagined. A timid person, fearful of embracing some activity wholeheartedly is often more accident prone. Psychokinesis, where the unconscious mind reacts to noxious stimuli to produce inhibition of muscle movement, has been attested to anecdotally by many reliable observers and has also been studied scientifically [193, p.221]. It was well-known that the scientist Wolfgang Pauli was able to unconsciously make scientific instruments in any laboratory he visited behave erratically through some unconscious telepathic process.

Freud was an atheist and regarded religion as a neurosis. He believed that people only turned to religion for a sense of security once they had left their parents and were alone in the world: God was an illusion that people created as a substitute father figure and imaginary protector. The father played the key role in the family according to Freud as boys identified with them in the development of their own ego, though often in the love-hate relationship of the Oedipus complex. But sociologist Nancy Chodorow, in her book *The Reproduction of Mothering* [49], attached much more importance to the mother in shaping adult gender roles for the child. Mothers usually have the key role practically in upbringing during childhood while the father is absent as breadwinner. Girls then model themselves on their mothers and carry the affection of the mother into adult behaviour. It is quite usual to see a woman hugging and discreetly kissing another woman as a greeting, and perhaps even a man whom she knows quite well. The growing boy defines his self by differences with the mother and will tend to distance himself physically from her as he develops. In many societies it is very rare for two men to hug when they meet – at most, they are likely to shake hands, if that. They may however welcome the opportunity for physical closeness by hugging and even discreetly kissing a woman they know. Chodorow suggests that this may be a dominant factor in producing the well-recognized male inexpressiveness – the difficulty men have in revealing their feelings to other people. These relationships are important since they define the roles we expect men and women to play in adult society: it is a moot point to what extent the roles of aggressive male and nurturing female are defined hormonally and to what extent they are socially prescribed [50]. In a study of more than two hundred societies anthropologist George Murdock found in most of them that the female was occupied with domestic and family responsibilities while the male worked outside the home, but with the woman often playing a prominent role in decision making [265]. Whatever qualities women bring to a situation it is clear that in a holistic society they should play a much more prominent part than they do at present, and not be merely subservient to men or given lesser remuneration or fewer promotion prospects.

Freud's conclusions are heavily criticized today because they were based on poor quality data that could not be verified as all diagnoses were subjective, the sample of patients was in no way representative of society as a whole, and there was no form of statistical analysis of the data. But his work was important because it helped society to think about women in a much more positive light, and he forced doctors to consider mental disorders in a more compassionate and much less mechanistic fashion. Psychologists acknowledge the importance of sexuality in providing motivation but do not attach as much importance to it as did Freud. This sexual drive, which Freud called the libido, was interpreted by later psychologists, like C.J. Jung, as a psychic motivating force but by no means exclusively sexual. French philosopher Henri Bergson (1859-1941) described the driving force of living organisms as the *élan vital* or life-force.

The most important aspect of Freud's work in a holistic approach to the world is to appreciate the significance of the unconscious. We know that we must be on our guard against being unconsciously brainwashed by advertising into continual material acquisition. It is our subconscious that must become imbued with a view of the inter-relatedness of all creation so that all of our actions are automatically and unconsciously directed towards re-affirming and building this relationship. It gives a theoretical underpinning to Pavlov's results on conditioning. Programming of our unconscious with thoughts of love and acceptance towards others would make our actions to this end second-nature. We know now from the work of biologists that the cellular environment, both within and outside the body, has a continual effect on how the DNA in our cells behaves [221,

222]. If we reaffirm messages of peace, love and health for ourselves and others, the cells become suffused with positivity and take us in the direction of wellbeing. If the deity is envisaged as a universal spiritual field with which the whole of Nature is imbued then the Freudian image of a guardian 'father-figure' may be relevant in that such a universal energy will interact with the natural world, and with humankind in particular, at every instant.

One of Freud's seminal publications was *The Interpretation of Dreams* (1900). Most of us pay little attention to our dreams, but they are of great importance in psychoanalysis as it is believed that they reflect the true feelings of the unconscious and they are also important in the mystical practice of Sufism (see next chapter).

The struggle for supremacy

This is a notion with which we are all familiar in today's society. In almost every aspect of our lives we seem to be engaged in a struggle for supremacy – especially in competition for jobs, for promotion in those jobs and for partners. Political and religious institutions have for generations been shaped by the quest for power. This concept was first described in a psychological context by Alfred Adler[4].

There are two key features of Adlerian psychology relevant to our discussion:

(i) First is the belief that while individual instincts and drives are important in determining the personality (the genetic or 'nature' element), the more important factor is the total social context in which the individual develops and how they respond to it (the environmental or 'nurture' aspect). If people do not 'fit in' with society and cannot feel at ease with their fellows then neuroses may develop. This view of the importance of the role of society may be compared with the societal *Geist* of G.W.F. Hegel influencing the wellbeing of the individual: if the individual is not in harmony with the societal philosophy, this will lead to *alienation.* Many contemporary psychologists and sociologists believe that our societal relationships are paramount in determining our success and happiness [122, p.78; 240, p.30]. Charles Hartshorne sees 'the social structure [as] the ultimate structure of all existence' [154, p.28] and even went as far as declaring that 'deity is the supreme case of the social principle'.

(ii) The predominant expression of the personal psychic energy of an individual is not in combating or repressing sexual desire, as suggested by Freud, but in a 'struggle for superiority' over their peers to assert their individuality – a milder version of the Nietzschean 'will-to-power'. As we shall see in Chapter 5, this reflects the views of Thomas Hobbes and the ancient Greek philosopher Thucydides. The 'will-to-power' can also be viewed in a positive and constructive sense as the determination to succeed, to make things happen for the better in the world at large, or 'the power of intention' as spiritualists call it [88; 168]. A certain degree of power in an individual is necessary if they are to influence positive changes in society.

In summary, where Freud emphasized the personal, individual and sexual background in development, Adler believed that it was the whole social environment from childhood onwards that shaped personality. Experience has shown that both factors contribute and their relative influence will vary from one individual to another. The quest for supremacy that Adler defined in the human personality is expressed in our world today through the economics of capitalism (see Chapter 6). Each man seeks, first, to keep up with his fellows in personal acquisitions, and then to establish pride of place though his display of power and material wealth. In western society, so often it is this that establishes the status of a man. In this regard, it is indeed society that is shaping the development of the individual in every way, as Adler suggested, but as Mary Midgley said, we can achieve more by cooperation than by competition [253].

Intelligence

The English explorer and anthropologist Sir Francis Galton (1822-1911), a cousin of Charles Darwin, turned his attention in later life to a study of heredity and intelligence. In 1883, he was the first to introduce the word-association test, much used by Freud and later psychoanalysts with their patients, and the first to devise a method for testing intelligence. Using Darwin's ideas of 'survival of the fittest', he suggested that the slow processes of 'natural selection' could be speeded up by selective breeding of those with the greatest intelligence, hence giving rise to the socially unacceptable principle of eugenics. As a result of this, Galton's name is not now held in the regard that perhaps it should be for his other innovative suggestions. He initiated the nature–nurture debate, and indeed coined the phrase, in his book *Hereditary Genius* of 1869, exactly a decade after Darwin's *Origin of Species* was published.

In France, in 1905, Alfred Binet and Théophile Simon developed their own tests for intelligence of children with special needs. The test was adopted, translated and adapted in the U.S. in 1916 by a Stanford University psychologist, Lewis Terman, to become the Stanford-Binet test, now used regularly in English-speaking countries. Terman was the one who devised the concept of the intelligence quotient, IQ.

In 1904, English psychologist Charles Spearman (1863-1945) suggested that there was a fundamental or general intelligence underlying all mental activities. To compare the intelligence of individuals at different kinds of tasks he devised a method of correlation between them by ranking their abilities. Spearman's Coefficient of Rank Correlation has become a standard measure of relative performance of two sets of variables in statistical mathematics. Since Spearman's day, psychologists have argued over whether intelligence is single or manifold. In 1938, Louis Thurstone (1887-1955) suggested there were seven 'primary mental abilities' comprising intelligence. In 1967, American J.P. Guilford came up with 120 intelligence factors, and by 1985 he had increased this to 150!

More manageably, Howard Gardner, in his 1983 book *Frames of Mind* [116], followed by the more recent *Multiple Intelligences* [117], suggested six kinds of intelligence – linguistic (ability to speak, read, and write languages), logical (ability for mathematics and problem solving), spatial (highly developed in people like artists, architects, pilots), musical (excelling in composers, musicians and those finely tuned as listeners), kinesthetic (ability for physical sports), and personal (both intrapersonal – an ability to know oneself; and interpersonal – an ability to relate successfully to others; these are sometimes distinguished as two separate intelligences). Personal intelligence has also been described as Emotional Intelligence, EI or EQ, a term that seems to have been coined by Peter Salovey [341; see also 128]. In 2000, physicist Danah Zohar and her psychologist husband Ian Marshall published their book on *Spiritual Intelligence* (SQ) [402], the ability for psychic awareness, not considered originally by Gardner or others before him, though Gardner did add an 'existential intelligence' to his original six or seven. The various intelligences are sometimes summarized by saying that IQ enables us to deal with the physical world, EQ enables us to relate to one another and to direct our passions, SQ enables us to relate to our inner selves and separate ourselves mentally from the material world around us, as in meditation. The development of these levels of intelligence, and particularly of emotional and spiritual intelligence, is clearly a priority of holistic existence. Our self-esteem and social integration depend in our early years on our IQ, our ability to learn to perform tasks (walking, talking, learning to read and write, to ride a bike, to play ball games, and so on) so that we can relate to other children and adults around us.

Another way of putting this would be to say that IQ is a function of mind, EQ represents that borderland between mind and soul and SQ is purely a function of soul. In this context, mind is regarded as the, in some sense, physical function of the brain and spinal cord – the function produced by the sum of the chemical and electromagnetic processes within the central nervous system. Mind is a function that we can to some extent measure quantitatively by the aforementioned intelligence tests. Soul is the aesthetic and moral component of mind function, a qualitative entity for which no quantitative measure can be given except in general comparative terms. We can say that the avatars and prophets, and great humanitarians like Mother Teresa, Mother Mirra and Sri Aurobindo have a more highly developed soul than the majority of humankind, and that most people, thankfully, do not have the sort of degenerate soul of murderers, thieves and rapists, but beyond this we cannot go in providing any quantitative measure of soul. Furthermore, soul is a function partly of the brain and partly of the universal spiritual field that I have described elsewhere as Communal Soul [193]. This is the component of human existence that survives mortal death to continue its existence within what spiritualists call the etheric plane, and this concept forms a most significant idea in a holistic world outlook. We are more likely to care for our fellows and the natural environment if we feel part of a spiritual dimension that transcends space and time to envelop the whole material cosmos.

The conflicts within us

The psychologist, Erich Fromm[5], developed Alfred Adler's idea of the impact of society on the individual. Because he was influenced by Karl Marx and G.W.F. Hegel, Fromm showed how the individual's role in life had been shaped by human history. In the early Middle Ages, our fate was determined purely by chance – we lived in whatever environment we were born into and choices of lifestyle were limited. Our fate in eternity was determined by God, and people, both nobility and peasantry, lived their lives under the authority of the Church. But with the Reformation, each individual was made responsible for the fate of their own soul, while the monarchies governed our earthly life. With the democratic French and American Revolutions we were given charge of the secular aspect of our lives too. The Industrial Revolution turned us from free self-employed farmers, craftsmen and artisans into employees working for the profit of others; we were no longer free to organize our days to suit ourselves but had to conform to our employers' schedules. The socialist revolution of Communism then made each individual responsible for the welfare of all others in the state. This was one responsibility too far, for the task was too daunting and beyond our control; individualism and incentives had disappeared in the state's need for conformity and control and, in the Soviet Union, which saw the birth of Communism, the system collapsed within a century.

These massive cultural and social changes within the space of some 500 years produced tension in our lives as individuals because we were alienated from the kind of lives we had become genetically and socially programmed to live. Fromm adapted this idea of alienation from Hegel and Marx. So the first of the internal conflicts of personality that Fromm envisaged was the struggle between individual free expression and the requirements of society, the Church, monarchy or state. As we saw in the 1990s, the power of individual freedom finally overcame nearly a century of state oppression when communism as a social and economic system failed in the Soviet Union and Eastern Europe. The opposite pole of social and economic conduct – capitalism and free enterprise – has generated its share of problems too. These problems are threatening to overwhelm humankind in the near future if we do not get them under control. Fromm put it starkly but succinctly three decades ago: 'the physical survival of the human race depends on a radical change of the human heart' [112, p.19].

In *The Fear of Freedom* [110] Fromm maintained that although we had thrown off the shackles of being bound to Church or state, because of our innate need to belong we now feel a need to look for new sources of attachment. 'This isolation is unbearable and the alternatives that [man] is confronted with are either to escape from the burden of this freedom into new dependencies and submission, or to advance to the full realization of positive freedom which is based on the uniqueness and individuality of man' [110, Foreword]. In our holistic vision of our lives we must decide quite early on – usually in our late teenage years or early twenties – to what extent we wish to be free agents and how great is our need for partner, family and responsibilities outside ourselves. If this kind of counselled discussion were part of every school syllabus, perhaps we would have more families in which children grow up in homes with their two birth parents. We must struggle within ourselves to resolve the conflict between the security we feel from a close, intimate and loving human relationship and the demands on our personal freedom that such a relationship requires of us, especially if there is a family of children or ailing parents involved. Psychologist Karen Horney (1885-1952) had a similar view of our feeling of anxiety at being alone or isolated from a community. We all need to feel loved or respected or wanted in some way and the unconscious thereby drives us to satisfy these needs in our social behaviour [176]. Fear of loneliness is not the only fear we have to confront in our lives.

The role of fear in shaping both our unconscious and conscious thoughts was referred to in the previous chapter. Love, anger and fear may produce similar physical effects in raising blood pressure and heart rate, and producing pallor or flushing, but the emotions feel very different to the individual perceiving them, and they provoke quite different physical actions. There are those who fear both success and failure to such an extent that they are almost paralyzed into inactivity. For success breeds its own fear, as there are always people waiting to knock you off your pedestal and replace you themselves: you set a standard that you fear you may not be able to sustain. There is a regrettable attitude prevalent amongst some schoolchildren in Britain and America to sneer at those who work hard or succeed at what they do by those who fear that they will never do so well. In America, the whole ethos of school or college is about striving to be the best, but fear of failure is never far away and the disadvantaged still mock those who succeed.

The 17th century philosopher and social reformer Thomas Hobbes (see Chapter 5) and the Greek philosopher Thucydides whose work he translated were both aware of an individual's fear of losing wealth, power and control, which was his principal motivation in always questing for more. Alfred Adler appreciated that our quest for supremacy was borne at least partly from this fear of losing power and control.

Fear of becoming a victim of crime makes people shelter behind locked doors or shut themselves away in fortified homes in every major American city and increasingly in Britain. The elderly have become fearful of walking about alone at night. Our fear of having our children abducted or having them maimed or killed in a traffic accident has made them fearful of ever being alone on the streets or even in open places and is turning them into overweight 'couch-potatoes'. Fifty years ago the elderly were less fearful and our children could walk and play happily in the streets or in open green areas without fear of traffic or abductors. Working to overcome this climate of fear for ourselves and our children is a key target of holistic living, for if we cannot liberate our children, what hope is there for the future? [see Sue Palmer's book on the subject, ref. no. 285]

In a comment on modern Western society Fromm perceptively remarks: 'One of the most outstanding psychological features of modern life is the fact that activities which are means to ends have more and more usurped the position of ends' [111, p.194]. Our working life, while it should provide us with some degree of satisfaction, is undertaken primarily to earn enough money to provide essentials and a few luxuries for ourselves and our families. Now, we rush about to earn money – often far more money than we need to live comfortably. We commute great distances, which is physically tiring, generally unproductive, and detrimental to the environment, so that we can get 'better' jobs – better in what way? Higher salaries perhaps, but also more responsibility, more stress, and probably less relaxation time or time to spend for oneself or with family and friends. The Industrial Revolution promised more leisure time by the introduction of machines; but all that happened was that mill and mine owners simply demanded more output for the same wages in the same length of working day so that they could make more profit and lead ever more comfortable and relaxed lives for themselves. The working men and women were rarely better off than before, and now had to endure tiring work in often unpleasant and uncomfortable surroundings to rigorous schedules. We have now invented computers and other electronic gadgetry to save time and labour. But again, either because we have to or because we want to, we tend to spend that time and effort that potentially we could have saved in rushing about making more money: 'We have become enmeshed in a world of means and have lost sight of ends' [*ibid.*, p.194].

Opportunities to increase individual leisure and pleasure promised by both industrial and electronic revolutions have not materialized. This is partly because of the expectations that society has imposed on the workforce in their having to adopt a certain quality of lifestyle: 'Pleasure, while being a subjective experience, cannot be judged in terms of the subjective element alone; it has an objective aspect, namely, that of man's physical and mental welfare' [*ibid.*, p.177]. It is also partly because of the demands of managers and directors for increased profit in commercial concerns, leading to longer hours and tired employees. Fromm paraphrases Thomas Aquinas in his *Summa Contra Gentiles* in saying that sin is not disobedience of some irrational authority, but the violation of human wellbeing [*ibid.*, p.122].

The more we can construct a living for ourselves within a limited geographical area, with less commuting and questing and more meaningful social and family relationships, the better our quality of life, and that of our children, is likely to be. Our existence has become too focused on material acquisitions (having) and too little on the spiritual dimension of life (being), both of which are essential components of fulfillment in bringing us true happiness or joy.

Fromm discusses this issue extensively in another book [112; see also193, p.202]. Fromm illustrates these different world-views lucidly by reference to three poems. The first of these is by Alfred Tennyson (1809-

1892) which illustrates the reductionist attitude of scientific investigation that has driven Western materialist philosophy for the past four centuries:

> Flower in a crannied wall,
> I pluck you out of the crannies,
> I hold you here, root and all, in my hand,
> Little flower – but *if* I could understand
> What you are, root and all, and all in all,
> I should know what God and man is.
> Tennyson, *Flower in the crannied wall*, 1869

The reductionist method is to destroy and dissect in order to understand and exploit Nature and thence know God. This is the materialist philosophy of 'having' that drives our society.

Fromm contrasts this with a haiku by the 17th century Japanese poet Basho. In an English translation, the haiku runs:

> When I look carefully
> I see the *nazuna* blooming
> By the hedge!

The poet observes and contemplates but does not destroy. He is content to embrace the flower spiritually and be at one with it – the philosophy of 'being'. We shall say more about this philosophy in the section on Taoism in the next chapter.

Finally, in his trilogy of poems, Fromm quotes *Found* by Wolfgang Goethe (1749-1832), a German poet, dramatist and scientist who lived to see the Industrial Revolution changing the character of existence:

> I walked in the woods
> All by myself,
> To seek nothing,

That was on my mind.
I saw in the shade
A little flower stand,
Bright like the stars
Like beautiful eyes.

I wanted to pluck it,
But it said sweetly:
Is it to wilt
That I must be broken?

I took it out
With all its roots,
Carried it to the garden
At the pretty housed.

And planted it again
In a quiet place;
Now it ever spreads
And blossoms forth.

Goethe is wandering without any specific purpose and comes across the flower. His first temptation is to pluck it, in the emerging materialist philosophy of his day, but then he decides it is better to transplant it so that it can generate more flowers of the same kind and then he can have a garden full of them to admire. Goethe's classic re-telling of the tale of *Faust* (1808, 1832) is similarly the account of a man willing to sacrifice his spiritual wellbeing in order to secure material ends. These three poems illustrate graphically the differences between a materialist 'having' attitude and a spiritual 'being' philosophy.

Being is the holistic attitude to life: 'Joy, then, is what we experience in the process of growing nearer to the goal of becoming ourself' [112, p.120]; 'Joy is not the ecstatic fire of the moment. Joy is the glow that accompanies being' [*ibid.*, p.118]. Joy is the goal of holistic living in a community of spiritual wellbeing. Fromm echoes the sentiments of the medieval German mystic, Meister Eckhart, when he says 'Man's task is not to retire into an atmosphere of spiritual egotism, remote from the affairs of the world, but to lead an active life in which one tries to contribute to the spiritual perfection of society' [*ibid.*, p.159].

Instead of always seeking material things to attain a superior social status, as Adler characterized the human condition, and which Fromm called the 'having' mentality, Fromm maintained we would be happier if we settled for those things that would allow us live comfortably and then concentrated on the quality of our lives – the 'being' realm. As another aspect of mind, schoolchildren could be encouraged to understand their subjects within a broader historical context (the 'being' state) rather than having to learn specific facts to pass examinations in order to get paper qualifications and boost school 'league tables' (the 'having' state). Within the spiritual realm, it is more important to do whatever we can to uplift the lives of others, that is, 'being' spiritual, rather than going though prescribed rituals to try to show others how dutiful we are in 'having' religious beliefs, but then acting selfishly, uncharitably or even maliciously towards other people. A vision of a divine spirit that is within us and all around us constantly interacting and motivating us in a holistic world (being) is more meaningful than a remote deity that we 'have' to turn to in times of need or who controls our lives from without (the 'having' mode). This self-preoccupation goes all the way up the social scale from the individual citizen to those we have elected to govern us, for they too 'value personal success [having] more highly than social responsibility [being]' [*ibid.*, p.19].

Although Fromm's books are perhaps not as widely read now as they were a few decades ago, they have much to tell us reflectively about how we might best achieve our way to holism in the present day.

The collective unconscious

Carl Gustav Jung (1875-1961) was born in the small village of Kesswil, on the shores of Lake Constance in Switzerland, but before Carl was four, the family moved to Basel where Jung's paternal grandfather, a respected physician, became Rector of Basel University. Jung's father was a Protestant pastor. His maternal grandfather was also a theologian but one who had a deep interest in the occult. His cousin, Hélène Preiswerk, was a medium and details of her seances were to form part of Jung's Doctoral thesis, *On the Psychology and Pathology of So-Called Occult Experiences*. Jung's family background of spiritualism, theology and medicine was to shape his future career path and his world-view. Because his mother was rather emotionally unstable, the family home was far from providing him with a happy relaxed atmosphere. This drove young Carl within himself, which was not necessarily a bad thing for this introverted, introspective personality served him well in adulthood in understanding the minds of others as he had been driven to understand himself. Because of his sensitivity he became subject to mental and emotional experiences that other people might never even have brought into conscious awareness. He was much more at home in the personal world of dreams, day-dreams and mystical visions than in the external world of real people.

Jung's first academic paper was presented to a student debating society at Basel University. It was called 'On the Limits of the Exact Sciences' and in this he criticized orthodox science for its inflexible materialism and reductionist approach. At the start of his professional career Jung worked with Freud for the six years from 1907 to 1913, but then differences emerged in their approach to analysis which demanded that they go their separate ways.

Jung did share the views of Freud and Adler about the importance of the unconscious, but his views were rather different. Jung, like Freud, accepted the importance of the unconscious mind in shaping the development of the personality but, expanding on the ideas of Freud, Jung suggested a four-fold division of functions of the mind. Jung defined the *preconscious* and *conscious* divisions of the mind in much the same way as Freud, but considered that the unconscious comprised two separate and distinct components. He defined the *personal unconscious*, which includes repressed wishes that are socially unacceptable and traumatic experiences, corresponding to the 'unconscious' of Freud's first analysis of mind function. But then Jung postulated another division of mind called the *collective unconscious* in which we share our mental experiences unconsciously with others. For consciousness itself is an energy field, a tiny fragment of Infinite Mind or the Cosmic Spiritual Field that comprises a boundless ocean of possibilities. It is the human mind that turns quantum waves of probability into the material particles of experience.

It was Jung's belief that we all have a predisposition to act in well defined ways under a particular set of circumstances. The reason we do this, he maintained, is because certain intuitive images are passed from each generation to the next, and even from one culture to another. These patterns of action or primordial images appeared in dreams or imaginings and found practical expression with recurrent themes found in myths, legends and rituals of many indigenous tribal groups whose spirituality had not been corrupted by technological advances. He called these patterns of behaviour *archetypes*, which he then defined as '*a priori*, inborn forms of 'intuition' . . . of perception and apprehension . . . Just as his instincts compel man to a specifically human mode of existence, so the archetypes force his ways of perception and apprehension into specifically human patterns' [194, para. 270]. The archetypes correspond to Plato's vision, presented in the dialogues *Phaedo* and *Meno*, of the Ideas or Forms that provide us with the templates for our conceptual notions. The realm in which the Ideas exist is effectively the same as the collective unconscious or Hegel's *Geist* or Sheldrake's morphic field, the spiritual medium through which the thoughts of one individual are transmitted through space and time to another. The archetypes also correspond to Richard Dawkins' cultural replicators or *memes* [76, p.206]. Prayer and distant healing, for example, have frequently been shown to produce a degree of healing in sick patients [193, p.231].

The simultaneous action of two or more people or occurrence of more than one event that we would interpret as meaningful coincidences Jung described as *synchronicity*. A similar phenomenon in which individuals acting independently find that their separate wills act in concert was described by the London-born futurist, Peter Russell, as *synergy* [321, p.95]. It may be speculated that any such interaction between the people or events involved, if such interaction does indeed occur, takes place by communication through the collective unconscious. Jung's patients often described some of these 'meaningful coincidences' that had occurred to them, but they had kept details of such events to themselves for fear of ridicule. The books by Alan Vaughan [374], Brian Inglis [185] and Plimmer and King [294] give many such accounts. It is quite impossible to assess such events scientifically because they are meaningful and significant only to the people involved and we have no way of knowing how many similar events occur that have no significance. However, when two people on their way to an appointment together meet in an unexpected place in a city of several million inhabitants, or other events with a statistically very small probability do occur and have special significance to the people concerned, one would be irrational in not considering the possibility that some spiritual agency might be involved. How often have we telephoned someone to have them say: "I was just going to call you"? Telepathic interactions between identical twins or other people who are emotionally close often fall into the category of comparable synchronous events involving transmission through the collective unconscious [293].

When societies work emphatically in concert to elect new governments, as in the British elections of 1979 and 1997, or they display an outpouring of communal grief, as in Britain after the death of Princess Diana in 1997, there is seemingly an element of Jung's collective unconscious or Hegel's *Geist* in operation created by unspoken communication between people's minds in addition to any rational assessment of the situation. The current rise in global spirituality was described by the American sociologist, James Redfield, in his Celestine books as arising from a similar shared communal ethos that he describes as synchronicity [308]. When stuck for ideas in the writing of his first book, *The Celestine Prophecy*, published in 1994, Redfield sought inspiration from the high-energy vortices of Sedona, Arizona, regarded by many as a mystical sacred site (see Chapter 4). The power of such sites resides in the spiritual energy they perpetuate through the collective unconscious.

Jung considered that the human propensities for religion and myth were universal expressions of these archetypes residing in the collective unconscious. Certain religious icons, like that of a divine redeemer son, may be found in Babylonian, Egyptian and even in Hindu and Mayan mythology as well as in Christianity [see my earlier book, *The Thoughtful Guide to God*, ref.no.193; and references therein]. Freud was very dismissive of religion regarding it as a 'universal obsessional neurosis' [108]. Jung, on the other hand, thought that religion was an essential social structure, and the shared rituals performed the function of allowing individuals to relate their inner feelings to the external world.

The collective unconscious then was made up of archetypes and therefore common to a whole group of individuals and perhaps, in some form to all individuals world-wide through time. Said Jung: 'Every individual life is at the same time the eternal life of the species' [195, para.146]. We are clearly shaped by both nature and nurture. Our genes may provide the predispositions to develop in a particular way, but our home environment and other external factors that we encounter on life's journey will determine the path our lives take. Modern biological research indicates that these life events feed back information to the genetic material in every cell of our bodies [222]. It is in our response to the environment that we exhibit our ability to resonate with the domain of the collective unconscious.

The development of an individual is recognised by psychologists to proceed in three general stages – from egocentric (covering the first seven years of our development from birth); to ethnocentric or sociocentric (where we develop deeper relations with our family than helpless dependence; and form friendships with our peer group in school and college – this lasts through adolescence); maturing finally into a worldcentric attitude where we are mature enough to embrace a global and holistic perspective. It is now as mature adults that our spirituality begins to develop; but the pressures of living in a developed society intrude. We may need to find a partner, perhaps one, perhaps many. We also need to find and develop a career and establish a home for ourselves and our dependents. The securing of these material needs comprises what Jung called the *acquisitive phase* of life. How long this lasts will be immensely variable, as some people seem to need to amass ever more wealth and power. Only when these drives are satisfied will we turn our attention to spirituality and the meaning of our lives that Jung described as the *inquisitive phase* of existence.

Jung believed that as we grow older we develop a realization of our full potential in the state that he called *individuation*, achieving the whole spiritually content individual as part of society and the natural environment. We have achieved our material goals during the acquisitive phase and found contentment and meaning in our lives in the inquisitive phase. As ecologist Thomas Berry said: 'Nothing is itself without everything else'. Individuation was the goal of life which Jung equated with wisdom. Aristotle's *eudaimonia* is usually translated as 'happiness', but it really comprises more than just that – it is happiness together with wisdom. For Jung, participation with others in the rituals of a formal religion allowed external expression of inner thoughts and feelings and comprised an essential part, for many people, of their psychological development and achievement of individuation.

Various psychologists and philosophers have seen the attainment of happiness as the goal of life, with happiness comprising the pursuit of pleasure, the avoidance of pain, and the attainment of satisfaction (self-respect, contentment and fulfillment). This last component Hegel called *Selbstbewusstsein* – best translated as being self-assured, having self-awareness, or being comfortable with being the person that we are – which Hegel, like Adler and Fromm, believed could develop only in a social context. These somewhat different notions of our ultimate goal in life, as opposed to intermediary goals *en route* to something else, reduce to essentially the same concept – that of holism achieved by providing happiness and fulfillment for ourselves and others through social contact.

Part of this process of individuation, said Jung, was admitting to ourselves, if we were male, that we had a gentler, more emotional, sensitive and compassionate side to our natures – qualities that we generally associate with our ideal feminine: Jung called this the *anima*. The corresponding quality in women, when they needed to be more aggressive or logical Jung called the *animus* – a reflection of how women thought the ideal man would behave. Women still occupy a subordinate role in many societies: it is more difficult for them to be promoted, especially to top jobs, and they are usually paid significantly less, even for doing the same job, despite gestures by politicians at establishing equality with men. We still have a long way to go to enable women to play a fulfilling role in our society, to their detriment and ours generally.

In our day-to-day life however we have to assume, to some extent, the kind of personality that is expected of us in a social or professional environment. This mask of personality that we present to the world Jung called the *persona*. If this is too far adrift of the kind of person we really are – if we cannot really 'be ourselves' – then psychological or emotional problems may arise. As Frieda Fordham says: our deepest human need is 'to relate the inner and outer man in equal degree' [102, p.71]. An essential part of this process of individuation is the realization of our 'brotherhood with all living things, even with inorganic matter and the cosmos itself' [*ibid.*, p.78]: as the ancient religions believed, 'all that exists lives'. Jung thought that the collective unconscious working through religion was one of the most important factors in helping us to secure this balanced state of selfhood, because it related us both to other people and to something beyond ourselves.

Jung also defined two kinds of personality types depending on the strength of their psychic drive or *libido* as he called it: Jung defined this as a general psychic energy not just sex drive as envisaged by Freud. *Introverts* draw energy into themselves from others and are focused on their own thoughts and feelings. They have limited social contacts and tend to be quite quiet, shy, cautious and retiring. They also tend to be uncomfortable with change. *Extraverts* direct their psychic energy out into the world. They are the performers, the gregarious ones, the 'movers and shakers' who get things done. In practice, all of us have a little of both within us – two sides to our personality. At some times we want to be left alone to reflect but at others we are happy to be the life and soul of the party; but most people tend to be predominantly of one type or the other. In general, those with an extravert personality tend to be happier and healthier than the introverts because they have a wider circle of social contacts and are not intimidated by change – which is all around us today. The extraverts are more resilient to stress while the introverts may go even deeper within themselves, more often to the point of clinical depression. Extraverts are more actively engaged in a number of pursuits, so that if any one fails it is not as significant an event as it would be to the introvert with only a limited number of contacts and interests. Extraverts have higher self-esteem while introverts often think poorly of themselves and are more likely to be hurt emotionally by unkind comments that the extravert will simply brush aside. In holistic living, we try to achieve a balance between extraversion and introversion, and between the *animus* and *anima*.

Jung's introversion was probably one reason for the personality clash that he had with his mentor, Sigmund Freud, who was certainly more of an extrovert. Another would have been that Freud, with his scientific training, was always looking backwards in time in the sense that he always sought causes and origins of things – a typical reductionist. Jung, on the other hand, looked forwards in that he was a dreamer who was more concerned with an individual's motives, goals and objectives (working towards individuation) and he envisaged some ideal unifying spiritual realm amongst humankind as a whole. The collective unconscious could be thought of as fulfilling just such a role.

As a spiritually oriented introvert personality, Jung had many psychic experiences himself and believed that many of his ideas came from his spirit guide that he called Philemon, an old man with a white beard and ornate outstretched arms, like wings of a kingfisher – the archetype of the wise old man. The winged image recalls the *Faravahar* icon central to the Zoroastrian religion. From these personal encounters, and the experiences described to him by Hélène Preiswerk, Jung had no doubt about the validity of psychic visions. He also had a lifelong belief in God, inculcated in boyhood no doubt by his elder family members and reinforced by his own experiences. When he was 68, Jung suffered a pulmonary embolism which nearly killed him. It was in this state that he had a near-death experience, with the usual concomitants of seeing oneself from afar (or, in Jung's case, the whole planet Earth from out in space) and a certain resentment at being brought back to real existence again. Happily, he recovered and lived to within a month of his 86th birthday.

Summary

This brief review of the lives of a few of the greatest psychologists of the 19th and 20th centuries and their seminal ideas points up some issues important in holistic living. The unconscious shapes all of our actions which, in turn, feed back to our mind-set. Our thoughts and beliefs become ingrained in the unconscious and thence affect us physically, so we need to use every opportunity to surround ourselves with wellbeing as far as we are able, to inculcate positive ideas of health, generosity and oneness with the world. Physical ill-health often has its origin in the disquiet of the unconscious. We need to be alert to the effect that advertising has in programming the unconscious to a materialistic world-view.

The quest for superiority defined by Adler is clearly a trait that we need to keep in check for the wellbeing of others. That does not mean we have to be demure and retiring, for we must achieve and retain some measure of power or authority if we are to be effective in persuading others of our holistic outlook. The time for disseminating this philosophy globally is getting ever shorter! The key is to use power wisely for the benefit of others rather than ourselves.

Knowing something about the inner motives and desires and conflicts that are common traits of humankind, pointed out by Fromm, enables us to search for these characteristics in ourselves and better shape our lives. Perhaps the most holistically relevant aspect of Fromm's psychology is the notion of 'being' spiritual in what we do in our everyday lives rather than 'having' ever more material acquisitions. We do not need to succumb to the brainwashing of the advertising industry and to always be seeking more material things to outshine our peers, if we tend what we have wisely and content ourselves with a simpler life. This does not mean living the austere life of an ascetic. Life is for enjoying and experiencing as much as we can safely without endangering others. Particularly we must overcome our fears and truly *live* our lives. Allowing fear to dominate us does make the unwelcome events more likely to happen. Worrying about them does nothing to resolve them. Difficult situations must be faced, assessed, decisions made and then left to resolve themselves.

Jung's concept of the animus and anima teaches us to value the masculine and feminine, the assertive and reflective, the rational and spiritual elements in us all. The collective unconscious gives us a notion of a spiritual realm to which we all can and do relate in all our thoughts. The collective unconscious is another in a long line of visions of the spiritual realm created by many philosophers, theologians and shamans, and by scientists as the quantum vacuum or zero point field. This is a gloriously unifying holistic concept.

CHAPTER 3

SPIRIT

The spiritual dimension of holistic living is represented by our faith, and that in turn is an expression of the feelings of the soul. The essential character of soul is connection with the universal spiritual energy, whether it is viewed as secular or divine. Many are content to develop their faith as individuals by introspection without the external support of organized religion. Others need the social context of religious rituals to sustain their faith, so the first part of this chapter considers spirit in the context of faith and religion with some of the different ideas about God. The laws that apply to the spiritual dimension are then explored in greater depth, moving on to a discussion of the principles of the eastern and western mystical philosophies.

I use a small 's' for 'spiritualism' to denote the practice of being spiritual, belief in an ultimately spiritual ground of existence that has influence over our lives. With the dire state of the planet it is certain that nothing but a universal belief in the sacredness of every action in our everyday lives, and in the sanctity of every life form on Earth, will enable humankind to survive into the 22nd century. A belief in a unifying spiritual cosmic energy sustained indigenous peoples for millennia and it is this that we must recapture, with or without organized religion. Regrettably, Christian missionaries for centuries have done their utmost to extinguish the simple holistic Nature spirituality of indigenous cultures and replace it with divisive Christian individualism based on man-made dogma. This is unfortunate because we have an urgent need at the present time to embrace a Nature spirituality.

Some spiritualists believe in a God, others in a God and Goddess, or a number of Nature gods, or none at all. They commune with spirits or angels, either spirits of the earthly departed or Nature spirits, and are guided by their wisdom. The spirits of human form are often those who spent their time on Earth amongst the indigenous peoples of India, Tibet or North America. Such beliefs are distinct from those of Spiritualism, with a capital 'S', the formal religion. This religion too has belief in the existence of an afterlife as the core of the faith, and Spiritualists regularly commune with and are guided by their contact with the spirits of the afterlife. For many in Britain and America, this religion is intimately bound up with Christianity. Spiritualism, the religion, if it is true to its founding principles, should not be aligned with any one religion. It should be focusing on the great deal of empirical evidence that there is for continued existence of the human soul in the afterlife. Like Unitarianism and the Quakers, Spiritualism should be open to believers of all faiths.

Adherents to western religions and many scientists constantly assert the lack of any spiritual dimension to the natural world. Yet the focus of religious belief is conduct in earthly life as a preparation for the continued spiritual existence of the afterlife. Thus in organized western religion, only humans have any spiritual component of being. In eastern religion, all living things have spirituality. In pagan belief, spirit is invested in all of Nature, including trees, rivers and mountains: this belief is variously called animism, panpsychism or hylozoism. In a holistic or ecological outlook, it is this universal spirituality that we need to cling to if we are to value the mystical contribution that birds, trees and rocks alike bring to our lives as well as to respect their moral right to existence. Even from a purely selfish humanistic standpoint, it is in our own interests to see that maximum biodiversity is maintained because such systems provide a healthier and more stable environment for humankind [396].

Although spiritualists live a life dedicated to the universal spirit, this is accomplished through service to others here on Earth in this life. They do not spend their lives in some irrational fantasy world. As the London-born mystic Henry Thomas Hamblin (1873-1958) said: 'It is impossible to be well or happy or to advance in the spiritual life without service and work and practical contact with the external world. We know that the external world is not real or lasting, in an absolute sense, but it is real enough to our present consciousness. This life has been given us for experience, therefore we must get all the experience possible out of life, and be as practical and helpful as we know how to be' [142, p.21].

Religion, faith and the nature of God

We live in times of religious conflict. Perhaps this has always been so, to a greater or lesser extent, even in Roman and Egyptian civilizations. Apart from a period of a few hundred years in the 8th to 11th centuries in Islamic Al-Andalus, when Jews, Christians and Muslims lived and worked together side-by-side in harmony [162; 252], the whole history of western religion has been one of conflict. The battles for control of the Islamic faith saw the centre of Islam based successively in Medina, Damascus and Baghdad. In the region dominated by Christianity, even in the early centuries after Jesus, there were conflicts with pagans, Arians, Donatists, Gnostics and other sects claiming to be authentic Christians. Then there were persecutions against Jews and later Muslims for many centuries thereafter, culminating in the Crusades and Inquisitions from the 11th to 15th centuries. For the Church of Rome there was animosity towards what became the Eastern Orthodox Church after the final schism in the 11th century, though the establishment of the Holy Roman Empire under Charlemagne in 800 CE did not help the cause of Christian unity. The Church had the Protestant rebellion and establishment of the Church of England to deal with in the 16th century, and then the deist movement, beginning in the 17th century after the publications of Galileo and Newton, that was spurred on in the 19th century by the discoveries of Darwin and Lyell [193, p.140 *et seq.*]. All of these were conflicts against the principles and the authority of the Church and they have spanned the previous two millennia.

Today, religious strife with its senseless killing of innocent people is still evident in everyday news – between Jews and Arabs, between Sunni and Shi'ite Muslims, and between extremist elements within Islam and the materialists of the West. This has produced a call from the more vociferous atheists to rid the world of religion entirely. But religion, with its underlying myths and traditions, fulfils an essential spiritual and psychological role in the lives of many, and it is not religion itself that is at fault. Moderate religion sustains and encourages the spiritual wellbeing of the vast majority of the world's population. But as many reject the authoritarian strictures of organized religion to pursue their faith in their own individual way, some fundamentalist adherents become more aggressive and intolerant of anything they perceive to be a threat to, or deviation from, or undermining of official authority. This is contrary to the spirit of holism just as the intolerance of atheists to all forms of religion is unholistic. It is a familiar theme; but the way of holism is to encourage tolerance of any form of devotion, as long as it leads to the wellbeing and uplift of the individual and those with whom he or she comes into contact in their daily lives. With expanding international travel and migration, and global communication through the Internet, rational people are now aware that there are very many paths to the divine and that no more credence can be given to one than another. The problem was perceptively and extensively discussed some years ago now by John Hick in *The Problems of Religious Pluralism* [167].

With the perceived threat to religious institutions, a number of those who want to defend their religious principles against what they see as the erosion of morality all around them have turned to fundamentalism as the most militant and extreme expression of their beliefs; but such entrenched positions do nothing to break down barriers between religions so that people of differing beliefs can live together in harmony. The hatred that burns so brightly in Northern Ireland and in the Middle East – to give two examples that continually confront us in the media – is the product of such intransigence. Sometimes the hatred and enmity is most passionate between groups who profess, broadly, the same religion – as with Protestants and Catholics in Northern Ireland, all of whom are supposed to be followers of the peace-loving Christ; or between Sunni and Shia

Muslims, all devotees of the prophet Muhammad, differing essentially in who should be in the positions of power and authority within Islam. For the past two thousand years, religion has fomented, encouraged and endorsed hatred and execution of those who did not accept some particular set of man-made precepts but who chose instead another way to follow what they believed to be divine will. Stated thus, one might reach the conclusion that religion itself is to blame for all the world's confrontations, but this is undoubtedly too simplistic a view, for territorial issues and personal quests for power also make a major contribution. It is not religion that is at fault but those who control it and use it as a means to power.

There is some criticism of religion however in encouraging irrational beliefs that do not contribute to holism. Some western religions regard women as subservient to men, even as corrupters and seducers of men after the myth of the Garden of Eden. As a result, women have never been allowed to play the fulfilling role that they should in the West. The situation in fundamentalist Islamic countries is certainly no better, and Orthodox Judaism is also male dominated. Perhaps in patriarchal societies men are also afraid of the power of women that comes with their wisdom.

There is the irrational view in some Christian sects that all infants are born evil. That they will err at times in their lives is almost certain, as normal human beings, but to taint them with evil from birth is totally counterproductive in producing law-abiding citizens and a holistic ethos. Ludwig Feuerbach (1804-1872) held the Christian religion responsible for producing in Man a feeling of alienation from self because people were continually being told by clerics that they were irredeemably sinful but for the grace of God [98].

The idea that almost any kind of wrongdoing can be forgiven just because a priest ordains it so with the admonition that the penitent must go away and perform some man-made ritual is also not conducive to social holism. Many of America's underworld figures trace their ancestry to Catholic Sicily and commit the most heinous crimes but are 'forgiven' as long as they are sure to attend 'Confession'.

The Catholic Church's prohibition of effective measures of birth control in an already desperately overpopulated world leads to the births of millions of unwanted children whose parents cannot afford to nurture them. In Latin America, the result is that death squads roam the cities shooting 'street children' and disposing of their bodies [153, p.98]. This is monstrously offensive in any sane society. Priests forced into celibacy but needing the warmth of human love turn to vulnerable children to satisfy their normal human desires. But still the papacy has the arrogance to claim that its edicts represent the word of God.

In a time of global communication and in a spirit of global cooperation, somehow the message must be imprinted into the minds of religious extremists that the world religions are simply different paths to the cosmic spirit – all are man-made social institutions for the sharing of spirituality, following the inspired words of a prophet or perpetuating age-old myths and traditions, but no one nearer to representing ultimate truth than any other. It has to be admitted however that the Eastern religious philosophies are generally more holistic and spiritual than those of the West, even when there is no concept of a deity as understood in western religion. Western religion is very much more authoritarian and prescriptive of belief and behaviour. Contemporary eastern religions are closer to the spiritual beliefs of indigenous peoples and those of the ancients.

There have been many wars and territorial disputes in the East over the past four millennia, so all has certainly not been uniformly peaceful there either. But all of their principal spiritual leaders, like Siddhartha Gautama, Lao Tzu, Kung Fu Tzu, and sages of the past century or so like Mahatma Gandhi, Sri Aurobindo, Sri Ramakrishna, and many others, have emphasized the oneness and spirituality of the world, the sanctity of all life (*ahimsa*), and the importance of letting each individual chart their own path to enlightenment. None has advocated the sort of indoctrination prevalent in western monotheistic religion, to follow the elaborated views of one prophet to the total exclusion of all others. The ability to find tranquillity through meditation and prayer within the busy world is a common characteristic of devotees of Eastern religious philosophies: Gujarati, a language widely spoken in India, does not even have a word to express the emotion that Westerners call 'excited'. Where the English language has over 2000 words to describe various forms of emotion, there are only 750 in Taiwanese and only 230 in Malay [199, p.319], indicating perhaps greater emotional tranquillity and harmony among these peoples.

Some form of spiritual faith is an essential component of holism; organized religion is not. It is perfectly possible to be spiritually content and whole within oneself, but many people in the West do find the source of their spiritual lives within formal religion. Theological writer and lecturer Ninian Smart (1927-2001) defined what he called the seven dimensions of any organized religious belief: (i) practical and ritual (Christian Communion; eating only *halal* meat in Islam, or *kasher* meat in Judaism; the ceremonies of the church, mosque, synagogue and temple); (ii) narrative or mythical (scripture); (iii) doctrinal (the *Dharma* in Buddhism; the Incarnation and Resurrection stories in Christianity); (iv) experiential or emotional (meditation, drawing spiritual energy inward to the self, mystical experiences, revelation); (v) social (Druids; Sangha; any religious community, like those in monasteries, convents or temples); (vi) ethical (prescriptions of individual behaviour; respect for all life and for the natural environment) (vii) material (the church, mosque, synagogue and temple; ceremonial objects; sacred objects and places in the natural world) [343].

Basically, religion is a man-made social institution that has evolved at particular times and places in human social history for a specific people. Scripture is derived either from the setting down of myths and traditions that have been transmitted orally for many generations (as in the *Torah* of Judaism, or the *Upanishads* and *Vedas* of Hinduism) or it has originated from the inspired vision of one man (as in Buddhism, Confucianism and

Islam). It always involves respect for some form of ultimate state of existence, or Ultimate Being, or God. The German theologian Hans Küng (b. 1928) has delineated three great religious systems: Hinduism and Buddhism from India, Confucianism and Taoism from China, and Judaism, Christianity, Islam and Zoroastrianism from the Middle East [209]. Indian religion focuses on the mystical and numinous, Chinese religion encourages followers to heed the words of the sages but to find their own way to enlightenment, while Middle Eastern religions demand obedience to the instructions of their prophets or, at least, the interpretation of those instructions by the elders of the religion.

The terms 'faith' and 'religion' are sometimes used synonymously, but they need to be distinguished. Faith may be defined as the internalized system of personal moral values and general code of conduct under which we live our lives, often taking our lead from the words of scripture. Faith is important because with every action, with every thought we create the universe anew at each instant. The word 'faith' is also used in a slightly different and often secular sense meaning 'trust' – as in faith in our doctors, or in a political party. Feinstein and Krippner [96, p.24] give a description in their book of 'personal myths' which would serve admirably as a definition of faith: 'personal myths explain one's world to oneself, guide personal development, provide social direction and address spiritual longings. They organize one's sense of reality and guide one's actions'. All of these qualities apply to faith too.

Contrary to what some atheists, and even some theists, maintain, faith is not incompatible with reason and may indeed be validated by it. My views differ from those of Alvin Plantinga and Anthony Kenny that 'it may be rational to accept a proposition though it is neither self-evident nor evident to the senses, nor held on the basis of any reason'. That to me is 'blind faith', an attitude of mind or world-view held for purely emotional reasons – like Freud's vision of belief in God as a surrogate father figure. The statement is also a distortion of language, for the definition of 'rational' is 'of the reason'. So it cannot be rational to hold any view if it is not held on the basis of reason. It is more logical to say, as Kenny also says in his book *What is Faith?*, that 'faith is by definition a belief in something for a particular reason', so that emotion and reason combine to shape our faith.

Faith is intuitive belief that surely should be supported by reason, though there are many who hold a faith who probably could not give you any reasons why they do so, usually because they were indoctrinated with it by parents or schools or society. But if such faith inspires them to lead lives committed to the welfare of others then surely there is no cause for criticism. I think however we would regard a faith as soundly based if there are reasons why it is held. If faith is a set of personal beliefs held without any rational or empirical basis it is hard to see how it differs from fantasy as many of its critics claim. If we have a secular faith – faith in our doctor or faith in the government, meaning that we respect their opinion or actions – for that I think we would all need reasons. I believe that religious faith should similarly be based on reason. As thinking beings we show contempt for the uniqueness of our human intelligence if we adhere to a religion the precepts of which conflict with our rational world-view. Reason does not exclude the possibility of an emotional or intuitive component to belief. Awareness of the spirituality of existence is one of the most important of holistic principles.

Using reason as well as intuitive interaction to support the notion of the cosmic spirit is an integral component of a holistic world-view. As pointed out by Hindmarsh [169], if religion does not take account of rational knowledge and continues to insist on tenets that directly conflict with the evidence of our senses and fundamental reason, then 'we deny the very God-given humanity we are at pains to validate.' According to Paul Davies [75], John Polkinghorne [297], Arthur Peacocke [287] and Alister McGrath, all eminent scientists, the application of rationality and the ideas that have emerged in physics and cosmology this century based on sensory observation tend to support rather than deny the existence of God. The important issues in a holistic world-view are that we can and should still use reason even in assessing mystical or spiritual events. Such experiences are not undermined by reason. True spiritual interaction will have a profound effect on an individual's behaviour, as reported by Happold [147] and James [188].

If we have faith in a religion it is surely because we believe rationally that the principles represented by the scripture of that religion accord with our world-view of how we should behave. Whether or not we regard any of the revelations presented in scripture as divine in origin is a matter of faith, as they clearly cannot all be unilaterally and unequivocally true. We have to make a decision as to which words of scripture most closely accord with the world-view that we hold rationally and emotionally.

It is important therefore that we distinguish between religion and the spirituality expressed by faith. As Brian Woodcock says:

> Religion draws lines. Spirituality reads between them. It tends to avoid definitions, boundaries and battles. It is inclusive and holistic. It crosses frontiers and makes connections.
>
> Brian Woodcock, from *This is the Day* (Iona Community),
> Neil Paynter (ed.), Wild Goose Publications, Glasgow, 2002.

This vision supports that of Ninian Smart when he says, 'no religion . . . contains the whole truth'; or to quote the German Jewish theologian, Martin Buber (1878-1965): 'Nothing is as apt to mask the face of God as religion'. It is spirituality that is the core of all faith – not religious dogma.

Holistic living encourages each individual to develop their faith, to find their own Way and to take from the different mystical revelations of East or West those aspects with which each can live most harmoniously – for self, for others and for the environment. We do not have to believe in or adhere to all the principles laid down in the dogma of any one formal religion.

Father Bede Griffiths (1906-1993) was a 20th century sage who, having been ordained into the Catholic Church, spent twenty years as a Benedictine monk, and then departed for India in 1955 to seek in his later life to explore what Eastern religion had to teach us in the West [136]. His message to us endorsing this holistic approach is that we should heed the spiritual philosophy of the East in our everyday lives and it is a timely one in our continual unabated pursuit of capitalist materialism.

It is therefore not organized religion that we should be turning to in the Way of holism but looking for truth and strength within ourselves. In a holistic, all-embracing view of the world, intolerant religious fundamentalism and belief in the literal truth of scripture are rationally and socially untenable. Fundamentalism is not even necessary to maintain and nourish a sense of spirituality and indeed we would be far better off without it. No fundamentalist sect can feel at one with any other group if both assert that their belief system unilaterally and unequivocally represents truth. In a holistic world we must all accept that there is a degree of truth in all mystical revelations and that they simply represent different paths to communion with the universal spiritual energy, secular or divine. As I have said elsewhere, which path we follow is at first a matter of chance, an accident of birth that governs the social group into which we are born. It is only when we are older that we have the maturity to seek an individual path for ourselves.

I am sure that many fundamentalists accept and value the progress of science in providing communication through telephones and computers, and the benefits of modern medical technology when they are ill; they drive to work, they fly away on holidays, and their lives would come to a standstill without electricity. But these same individuals so often reject established scientific truths when it comes to phenomena in the natural world if they conflict with biblical accounts, where religion is presuming to be authoritative in a domain of experience of which it has no knowledge. The purview of religion is to counsel on matters of morality: it can have nothing meaningful to say about empirical discoveries and their interpretation in the coherent, rational system of science.

Ultimately, our connection with the spiritual domain has to be personal and individual, even if we adhere to one of the western orthodox religions. This is even more especially true of the peoples within native tribes in less sophisticated non-technological societies. The spirit guides of western healers and spiritualists almost always originate from ancient indigenous cultures in Tibet, China or India or from North American Indian cultures. Finding the way to holistic living is ultimately always an individual life choice for which we have to find the strength within ourselves. To quote the North American sage White Eagle: 'The secret of strength lies in the quiet mind . . . You will receive help as you ask, not with your mind, but in the depths of your being, quietly'. White Eagle's vision of universal love and spiritual strength has been transmitted to us through the medium Grace Cooke in her book *Meditation* [61] and other publications of the White Eagle Trust [385, 386]. On the same theme, I have used before a quote from French philosopher Blaise Pascal (1623-1662): 'the sole cause of man's unhappiness is that he does not know how to stay quietly in his room' [286, No.136 in the Krailsheimer edn.]. This accords with a prevailing eastern view of happiness as being derived from rejection of materialist questing to look for serenity within by contemplation.

Images of the Divine

Primitive people saw gods in many of the features of the natural world around them. As civilizations developed, these god-like qualities were transferred to real people, though the nature gods continued and imaginary figures were conceived to represent them. Ruling kings were regarded as god-like by the ancient Babylonians and Egyptians. All the Pharaohs were treated as gods. The Roman Emperors were equally seen as awesome, intimidating figures of god-like power of life and death. It is only a few decades since the Japanese Emperor ceased to be venerated in the same way, and the infallibility conferred by believers on the Dalai Lama or the Pope is a god-like quality.

The 19th century vision of God as being, in some sense, 'out there' acting as a supervisor of our earthly life has now been abandoned by many non-fundamentalist adherents to western religion. There have even been several eminent theologians who have gone so far as to lay out a new non-realist vision of deity. In his Foreword to the edited volume *God and Reality* [67], Rowan Williams, then Bishop of Monmouth and now Archbishop of Canterbury, questions whether a 'non-realist' view of God has any significance. Does the concept of God make any meaningful contribution to humankind if it is defined as: 'God is the imagined ideal unity of values' [*ibid.*, p.vi]. Much more meaningful, but now no longer non-realist, is his next suggested definition: 'God is a spirit of infinite power, wisdom and goodness' [*ibid*]. The latter accords with the holistic vision of the Divine for theists, but secular spiritualists have a similar vision of the power, wisdom and goodness of the interactive spiritual energy that embraces the souls of the living and the earthly departed. Spiritualists have always believed that the souls of those who are now in the afterlife have access to a wisdom that only relatively few of us can share during earthly life.

The non-realist view is also expressed in many of the books by Don Cupitt [68, 69]. Cupitt is an Anglican priest and lecturer in the Philosophy of Religion at Cambridge University. In *God and Reality* he states: 'The non-realist Christian is a postmodern person who has lost all the old beliefs about finding the perfect, normative world elsewhere, either in the past, or in the world above, or in the future . . . There is only one world, the world that we have built up around ourselves, the world produced by our language. [*ibid.*, p.21]. So 'God' is merely a linguistic label that we attach to some imaginary paragon of virtue, like Plato's Idea of the Good. I doubt that such an image would have inspired the great works of art, music and architecture that have been created as homage to the deity. God is surely so much more than what I have described as 'faith', though faith is the practical expression of soul, that tiny part of the cosmic spirit within us.

Cupitt thinks 'we don't need any higher Purpose' [69, p.91] and 'there is no point in our having any ineffable private experiences that can mean nothing to anyone else' [69, p.46]. How wrong he is! For most of us, those 'ineffable private experiences' and the 'higher purpose' are socially indispensable and a fundamental quality of what it is to be human. Individually, the 'higher Purpose' is a profound source of inspiration, and communally they bind us locally and globally. The Anglican cleric Anthony Freeman also aligns himself with the non-realist camp when he says that, for him, God 'has no independent existence' and he describes God as 'the sum of all my values and ideals in life' [67, p.26; 106, p.25].

The idea of God as within the mental or spiritual part of the living world, instead of being in any sense out there in physical space, was also expressed by marine biologist Alister Hardy: 'I believe that the nature of God is essentially one of personal qualities and that man's relationship to this presence must be a devotional, personal I-Thou feeling' [149, p.136]. If God is viewed as universal spirit then it is certainly within all of us and everything, animate and inanimate, but as a real entity. The uniquely human quality is that we can respond to it. The concept of a universal spirituality is a central feature of holism, whether or not we regard that spirituality as possessing qualities that we associate with the Divine.

The God of Freeman and Cupitt represents the non-realist position – the term 'God' is little more than just a linguistic device. It is however a great deal more than this because it has a socially binding symbolic moral value. Even Richard Dawkins from his atheistic position acknowledges that the God meme (a *meme* is a cultural replicator) 'has great psychological appeal'. . . it has produced 'great music and great art'. . . 'it provides a superficially plausible answer to deep and troubling questions about existence' [76, p.207]. 'Plausible', yes, but not superficially so, given the huge amount of both anecdotal and scientifically gathered evidence on psychic and spiritual phenomena or psi [36; 99; 193; 348; 383; 395; 399], supported, as they are now, by interpretation in terms of twentieth century quantum physics [193, p.277]. As I have said elsewhere, there is no reason why we should not believe the visions of mystics just as we accept the visions of scientists and mathematicians who tell us some incredible facts about the world that most of us are in no position to directly verify. They must be verified by others schooled in that particular world-view – just like the precepts of religion.

Daphne Hampson, Senior Lecturer in Divinity at the University of St. Andrews in Scotland, writing in *God and Reality* [67, p.85] presents the position of the non-Christian realist. As she says, if one does not believe in the real existence of God, and thence that God is incarnate in Jesus, surely one has no grounds to call oneself a Christian at all, whether non-realist or not? 'One can only hold to such a uniqueness through a faith which flies in the face of reason' [*ibid.*, p.86]. A further charge against Christianity in Hampson's view is that it is 'ethically untenable' and that it 'has been the undoing of woman in Western culture' [*ibid.*, p.87]. She points out that women have always been viewed by the Church as temptresses and seducers: 'in a period of about 200 years from the 1490s . . . something like a million women were burnt as witches'. Human sexuality and 'occult' practices like spiritual healing and divination have always been anathema to the Church. Daphne Sampson, on the other hand, has 'not the slightest doubt as to the powers of healing, extra-sensory perception and love' as real and meaningful spiritual communication between people.

Amongst those who accept the Divine as an objectively real being, or as a meaningful interactive spiritual energy, is Bishop John Robinson to whom 'the word 'God' denotes the ultimate depth of all our being, the creative ground and meaning of all our existence' [319, p.47]. He says further: 'our experience of God is distinctively and characteristically an awareness of the transcendent, the numinous, the unconditional' [319, p.52]. This, and Rowan Williams' second definition of God, are close to the holistic spiritualist view: God is not 'real' in the sense that it can be experienced with the five senses, as demanded by the logical positivist view of reality, but it is meaningfully, objectively existent and accessible to the sixth sense of mystical insight. This sixth sense is the activity of soul, distinguished earlier as spiritual intelligence, SQ. As I have said elsewhere [193, p.147], the notion of soul is also more than just a useful linguistic term to describe an aspect of mind or personality. It represents an integral part of human personality that can be differentiated from those experiences of mind that are in some sense quantifiable in IQ tests. Mystical experiences need not involve a feeling of being in the presence of the Divine [*ibid.*, p.169]; awareness of the inspiration of the universal spiritual energy in nature, music and poetry may also be regarded as mystical. Such experiences form an essential part of holistic existence and the achievement of individual fulfilment or individuation as Jung called it. Holism requires only an awareness of an underlying spiritual cosmic unity, whether secular or divine.

There are basically three approaches to authentication of the reality of this cosmic spirituality. The first of these spiritual paths is represented by individual revelation through meditation, prayer, or contemplation in the hope of gaining direct personal mystical experience. The insights given to prophets, seers, shamans and scribes in producing the works of scripture are of this kind, illuminating paths to personal and social holism. Meditation or prayer may be regarded as a channel between the personal consciousness and the collective unconscious. But true and deep introspection and awareness of ultimate reality is given to relatively few of us, even if we are practiced at losing ourselves in the numinous through meditation. It is usually confined to shamans, sages and prophets. A less demanding and more accessible form of spiritual awareness, that elsewhere I have called 'incipient mysticism' [193, p.169], is to immerse ourselves in natural beauty, or a work of art, poetry or music as an expression of soul.

The direct spiritual awareness that each individual and each social group has in indigenous societies, and which is still the basis of pagan belief, has been replaced in much of western society by the bureaucracy of organized religion. This has necessitated the establishment of intermediaries, 'experts', who claim to know how the message of the Divine should be interpreted. In fact, they have largely confused and obscured direct communion with the spiritual realm. They have distorted the real purpose of human life of holistic existence devoted to our fellow man and to the global environment and replaced it with obligatory man-made rituals and subservience to a powerful religious hierarchy. Indigenous pagan societies created their own religions based on mystical insight. Contemporary neopagans may be mono-, bi-, poly- or atheistic, but they all have a belief in a fundamental underlying spiritual unity. Living holistically in communion with this universal spirit is what binds humankind together and makes us an integral part of our environment.

Another path to spiritual uplift is that of prophetic or scriptural revelation, where we accept and trust the words of prophets, seers, mediums and mystics who have refined this power of introspection to such an extent that they can access truths that are beyond the reach of most of us. The process of revelation is the same as that elaborated in the previous paragraphs, but now we are relying on the mystical insight of those who are gifted and experienced in such techniques. The process is no different to our acceptance of the words of scientists and mathematicians as to their findings and their interpretation. The scriptures give us highly edited accounts of some of these divine encounters and the way of life they recommend (or what we are told they recommend!).

The third approach to establishing the existence of a universal creative spirit is that of so-called 'natural theology' – using our five senses and powers of reason. It is more holistic to use reason together with intuition rather than relying solely on our own emotions or the intuitive insight of the prophets because the interpretation of any kind of subjective emotion is not verifiable by others, and acceptance is a matter of faith.

A major difference between Western religion and the Eastern belief systems is that in the East, individuals are taught to look for truth and the solution to life's challenges within themselves, with guidance only from elders and sages. In the East, people retain the awareness that indigenous peoples have of closeness to the Earth and spiritual unity. In the West, for two millennia at least, people have been taught from earliest age to do and to believe what they are told by priests, imams or the pope, thus stifling freedom of independent thought and action.

> Find enlightenment through heeding many points of view
> Find ignorance through heeding few
> Wei Zheng, Tang Dynasty

In several major eastern religions, such as Buddhism, Confucianism and Taoism, there is no overriding God concept to impose some code of behaviour supposedly advocated by scripture or divine authority. True faith in the East is a belief in the universal spirit that imbues the sunlight on the morning dew or the mist that cloaks the mountain tops with an aura that transcends space and time, beliefs and traditions, cultures and customs. For all their lack of belief in a God that resembles anything in western religion, it must be said that the eastern mystical traditions are far closer to this one-world spiritual philosophy than the western religions with their constant enmity and hatred engendered by their man-made doctrines and dogma.

At various times during the period 800-300 BCE there was a huge change in human consciousness that affected much of the civilized world at the time and which has left an indelible mark on humankind ever since. This is the period that the German philosopher and psychiatrist Karl Jaspers (1883-1969) called the Axial Age because of its importance in human cultural development [190]. In the Middle East there arose a series of prophets and philosophers with ideas of the Divine and how humankind should live in accord with the principle of spirituality – Zoroaster in Persia; and Micaiah, Ezra, Ezekiel, Hosea, Jeremiah, Isaiah, and Nehemiah in Mesopotamia among the Jewish people. The philosophers Socrates, Plato and Aristotle emerged in Greece some four centuries before Jesus and they expanded the rational ideas of the Pre-Socratics like

Anaximander, Anaximenes and Heraclitus about the nature of the world and the qualities of the good life. This rationalism was complementary to the spiritual insight of the sages Siddharta Gautama (the Buddha), Lao-Tzu and Kung-Fu Tzu (Confucius) in China; and the authors of the *Upanishads* in India.

This degree of synchronicity in the emergence of an underlying spiritual base to human existence is most unlikely to have occurred by chance, and a degree of spiritual resonance achieved through what Jung called the collective unconscious or what Rupert Sheldrake denotes as the morphic field seems a reasonable speculative hypothesis. Many contemporary sages believe that such a spiritual revolution is again underway at present, as discussed above – indeed, that it is essential if humankind is not to destroy itself with the hatred and intolerance generated by western religions competing for unachievable world domination. Christianity has achieved its predominant position amongst the religions of the world not by presenting any greater degree of veracity in its dogma but only by ruthlessly and systematically exterminating anyone who refused to accept its man-made precepts for hundreds of years. Islamic fundamentalists are now attempting to pursue a comparable philosophy. But the many religions of the world are now well dispersed in their adherents. We are now all aware of the many different paths to the Divine, and able to exchange views peaceably with others of different beliefs, so that no such religious or racial persecutions must ever be tolerated again in human social history.

The principles of spirituality

Between
Our birth and death we may touch understanding
As a moth brushes a window with its wings
The Boy With A Cart, 1945

We must each find our separate meaning
In the persuasion of our days
Until we meet in the meaning of the world
The Firstborn, 1946
Christopher Fry (1907-2005)

There are certain over-arching principles that govern the way spirituality works in the world. The ultimate purpose of our lives is a spiritual one. We cannot exist without the physical dimension of nutrition and exercise. We cannot function successfully without stimulation for the mind and recreation. But the most important aspect of all in holistic living is the spiritual, for it is this which gives purpose to our existence and tells us the direction in which we must travel – and it is eternal.

In this section, we turn to the constructive and uplifting topic of the principles of spirituality that can lead to a holistic existence for individuals, and thence for families and local social groups, and then . . .the world! The organized religions have their tenets expressing the traditional observances required of the faithful. The Jews, for example, have their code of commandments or obligations (*mitzvoth*) drawn from their scripture, the *Torah*, and the Ten Commandments of Exodus [Chap. 20] are the fundamental principles on which the western systems of law and religion are based.

There are principles or laws of spirituality that govern enhancement of spiritual behaviour among humankind to promote holism just as there are so-called laws of physics that govern the behaviour of the natural world. The laws of spirituality are not verifiable to order like the laws of science but rather their truth may be ascertained in the manner prescribed by pragmatists like Charles Pierce and William James – by the positive effects they have on individual and social welfare [189].

Many writers describing a spiritual way of life refer to these universal codes of practice as Laws of Spirituality. But any rationalist seeing the word 'laws' will assume that we are dealing with concepts that are accepted as invariant by all of spiritual outlook, just like the laws of thermodynamics in science, or the Jewish *mitzvoth* – but this is not the case. Thus, for example, a popular contemporary writer on spiritual themes, Deepak Chopra, gives us seven 'Laws' [51], where he equates laws of spirituality with laws of success; Dan Millman gives us twelve [259]; while Diana Cooper gives us thirty-six in her book, *A Little Light on the Spiritual Laws* [62]. Pierre Pradervand describes four or five in his treatment of the subject [302]. So the classification of these concepts is largely a matter of personal outlook. This is not to imply that 'more' is 'better'– the same essential messages may simply be interpreted and presented in different ways, or different writers may see a need for different priorities. They are representations of the fundamentals of spirituality that underlie our earthly existence and which can serve as guidelines towards a holistic spiritual life. What follows is an amalgam of these ideas from many sources, including those mentioned above together with some thoughts of my own. Some of these themes of spirituality – unity, balance, attraction or *karma*, reflection or meditation, action or *dhamma* – will inevitably overlap in a holistic world-view.

The principle of unity. The various religious philosophies of the East differ in many of their beliefs, but the concept of unity is so universal that we probably could call it a Law, as it is also a belief of theosophists, pagans and others seeking spiritual truth. They differ in how this unity is expressed in the material world. Of the western religions, the Quakers, Baha'i and Unitarians are probably those that remain closest to the principle of unity. The view of God as a being that is utterly transcendent and separate from the created world is a characteristic of western religion's dualist position. The Cartesian separation of mind or spirit from body or matter has defined the path of western science for four centuries. In order to 'bring God back to Earth', some theologians like the German mystic Meister Eckhart (*ca.*1260 – *ca.*1328) proposed an utterly transcendent Godhead as Creator about which nothing can be said other than it created a God who was interactive in the world of humankind.

Going still further, the monist pantheistic assertion of philosophers like Baruch Spinoza (1632-1677) sees all of the created world as God: 'Except God no substance can be granted or conceived' [353, Proposition XIV]: all that is, is God; and 'God is not only the effecting cause of the existence of things, but also of their essence' [*ibid.*, Proposition XXV]: God is not only the Creator of all that is but is also indwelling. So the unity of the material and the divine is to be found in western theology and philosophy too.

In the East, the created world is regarded as *maya*, usually translated as 'illusion', though this implies that material objects do not exist. That is not the intention. If the world were fantasy or delusion it would be called *moha*. What it does mean is that all we can really know of the world is what we perceive with our minds and those visions form a continuum with the spiritual domain: as an Indian sage declared: 'you realize beyond all trace of doubt that the world is in you and not you in the world'. The 'unity' comes from the vision of a monistic spirituality within the human mind continuous with that beyond in the universal spiritual field. This unity is the central dogma of that division of Hinduism first specifically elaborated by Sankara called Advaita Vedanta. Put another way, consciousness pervades the entire natural world. From an ultimate spiritual perspective therefore, in this sense, there is no differentiation within the objects of the material world: all is one. Maya also implies that the existence of the physical world is a manifestation of the spiritual unity, a version of Spinoza's pantheism and David Bohm's 'unbroken wholeness', the explicate manifestation of the implicate order [31].

In quantum physics, since all matter, both living and inorganic, is composed of subatomic particles, and these engage continually in a cosmic dance with the energy field, all creation is a continuum: which is why physicists like Fritjof Capra, and authors like Gary Zukav, talked of the similarities between 20th century physics and Eastern mysticism [44, 403]. From a humanist viewpoint, there is no differentiation between people on the basis of race or creed – all humankind is one. That oneness is divine, which resides in us as soul. Although we are all different as individuals, we are like the snowflakes in a snowstorm that melt down to become part of the ocean of the Divine. We are part of that collective unconscious or Communal Soul finding expression in different bodies. Those bodies are made of molecules, the molecules are made of atoms, the atoms of protons, electrons and neutrons, and these subatomic particles are made of quarks and gluons, which are made up of . . . an ocean of pure cosmic energy, to use our holon concept. If physicists can regard such an energy as spontaneously creating the universe, then theologians are entitled to envisage that same field, now wearing the mantle of the Divine, as Creator.

The principle of balance. This is another way of restating the principle of the ancient Greeks of finding the middle way, of avoiding extremes. It is the substance of the well-known homily: variety is the spice of life but moderation is the essence of life. The ancient Greek philosopher Heraclitus said that all is change: the world is made up of a balance or equilibrium between opposites. In holistic living we need to resonate with that equilibrium. This is such an important principle for those who are too focused on their work, devoting too little time and attention to relaxation, family and friends. It's an important principle to be borne in mind by those who eat too much to maintain a healthy body weight, or exercise too little, or who drink too much alcohol so that they lose control and inflict harm on others or damage themselves. Food, work, exercise, sleep, relaxation – all need to be kept in balance in holistic living. When we are out of balance, our mind and our body will tell us if we but listen.

The principle of balance is an important one that scientists and religious fundamentalists should note. A religion will not be respected in a world shaped by science and technology if its fundamental tenets ignore or contradict the most basic of rational, scientific principles. On the other hand, scientists do a disservice to their profession if they insist that empiricism and rationalism constitute the only route to knowledge of how the world works, ignoring the inspirational and intuitive insights given to sages for millennia, that shaped successful societies long before science had begun. A holistic world-view demands a balance between empiricism, rationalism and mysticism.

The principle of balance is an essential component of the Taoist belief that the human personality should display an equilibrium between the feminine, intuitive, contemplative element, *yin*, and the masculine active rational principle, *yang*. This idea found expression in Jung's concept of the mix of animus and anima, male and female elements within the personality. There are times when we need dynamism, we need to be proactive, but at others we should just sit back and let events take their course. Establishing balance in the way we live is another element of holism.

The principle of attraction, which in some instances becomes *the principle of karma* or *principle of choices* or *principle of reflection.* Essentially, we tend to attract to us whatever we make the spiritual focus of our being. If we are gentle and loving to others, we reflect that love and affection back to ourselves – we must choose the path of peace and love. It is an expression of the power of intention [88, 168]. On the other hand, if we wish harm to others, are aggressive, or full of hate, we will be treated thus in return: 'whatsoever a man soweth, that shall he also reap' [St. Paul's Letter to the Galatians, 6. 7]. We attract to ourselves the intent of the thoughts in our conscious and unconscious mental processes. The chemicals in our body that control our emotions are addictive. If we find love or hate exhilarating, without conscious control we will seek more of the same. If thoughts can influence water, as the Japanese scientist, Masaru Emoto, has shown, our thoughts can certainly influence other living beings that are largely made up of water, as claimed by homeopaths.

This principle is all about using our faculty of free will wisely. The more beneficence we dispense and disperse into the aether, the more others are likely to pick up on loving thoughts and to treat other people in the same way themselves. Put another way, by our nature and intent we tune in to loving or hostile frequencies in the universal spiritual energy and establish rapport with like minds. It is well established that children who misbehave are often copying the behaviour of their parents or peers. Women often come out of one abusive relationship only to find themselves in another, unconsciously seeking a partner who will reinforce their feelings of a lack of self worth.

The principle of reflection is about treating people as we would wish to be treated ourselves in accord with the Categorical Imperative of German philosopher Immanuel Kant (1724-1804): "I will be constrained by reason to act only on that maxim which I can at the same time will as a universal law". We behave with goodwill to others hoping that they will treat us in the same way. As Mary Midgley says, cooperation produces better results more quickly and more often than competition. We should not be taking the Darwinian view of evolution – the survival of the fittest, as Herbert Spencer called it – as our 'religion' or model for an ideal way to live [253]. There is good reason why Deepak Chopra associates laws of spirituality with laws of success. We are likely to get ahead in our careers more by helping people than by ruthless competition. In practice however, in the capitalist economic system, it is the aggressive and ambitious employee willing to trample on the feelings of others who too often succeeds. Having always the intention to succeed, but to do good for others as well as yourself, will bring far more rewards, if only spiritual ones, than being ruthlessly ambitious. Those who trample on others on the way up can expect to be treated likewise on the way down.

Although in physics we learn that 'opposites attract' (and sometimes in life too – for excitement perhaps, or because we want a partner who complements our own personality?), in spirituality and social contexts the rule is 'like attracts like'. If we go through life with an air of acceptance and optimism, we find perhaps unexpectedly that things do work out more often than not and other optimistic people like having us around. If on the other hand we always fear the worst, it also usually happens! We have all met people whose goodness and joy of living seems to radiate from them and, conversely, others of whom we are instantly wary because their aura seems to send out negative vibrations that they are pessimistic in outlook or may not be trustworthy. People who are happy and successful are usually optimistic. Those who live their lives in fear, of any kind, are more timid and less adventurous and are likely to be in poorer health. As we found in earlier chapters, it is the extrovert who is happier, healthier and more beneficent while the introvert has fewer friends and is often not in as good a position to distribute (or receive!) largesse. When ill-fortune strikes, a holistic individual will regard it as a learning experience and turn it to advantage: as Pradervand says, 'Trials are blessings in disguise, and hosts of angels follow in their path' [302].

Many spiritual healers find that the effectiveness of their healing intensifies as they continue to practice in the same place. While physical energy is diminished by use, psychic energy is increased by use as the resonance with the universal spiritual energy intensifies. If several healers are practicing together, this may increase the power of their healing – though not always, for the respective healing energies may not be compatible. Spiritual healers are often known as 'channelers' because their aim is to establish a connection with the universal spiritual field and let the spiritual healing energies flow through them to the patient to do their work stimulating the body's natural healing processes. People who are considered to be holy are often depicted in paintings with a halo around their heads. This represents the aura emanating from the crown *chakra*. As described in Chapter 1, the chakras serve as a link between the aura and the physical body. Rationalists may not accept this concept but then they also dismiss the concept of meridians of *ch'i* in the body that acupuncturists have tapped into to produce healing or anaesthesia for many generations. By connecting with the universal *ch'i* healers attract to themselves, to pass on to patients, the universal energy to work a kind of spiritual alchemy, turning physical or emotional sickness into health and wholeness. In all areas of our lives, when we give love, we receive it in return.

There is scientific experimental evidence to suggest that people who are ill and who are prayed for by others unknown to them have fewer symptoms and recover more quickly than the unfortunate patients who act as controls who were not prayed for [39; 83-85; 152; 337]. A positive outlook and a relaxed temperament have many measurable positive effects on health as described in Chapter 1. Conversely, and again measurably, people who are short tempered and angry have higher blood pressure and pulse rate and are more likely to suffer illness. Stress on the nervous system and susceptibility to infection because of a weakened immune system are inextricably linked.

Though the Victorian idea of eternal punishment in hellfire for earthly transgressions has thankfully long been abandoned in Christianity, the effect of *karma* is to make us feel a sense of regret in the spiritual realm, if we do not feel it on earth, for a life lived selfishly without sufficient regard for others. The influence of *karma* is to attract to ourselves those experiences we need in order to learn this concept of living for others as well as ourselves. If we have ill-treated others in one lifetime, we may need to live a life on Earth experiencing the anguish or difficulties that we inflicted on others previously. This concept of *karma* is fundamental to Eastern religious philosophy and is viewed not as a punitive but as a learning process.

Many of those who believe in reincarnation maintain that we are sometimes reincarnated to live amongst those to whom we were closest in a previous life. Several psychologists believe that problems encountered in a past life may be responsible for certain emotional imbalances in this one, and that these disturbances can be resolved by techniques such as past life regression under hypnosis [99; 383; 399]. The evidence provided by subjects on so many occasions, much of which could be authenticated, is far too detailed to be imagined and such a large number of cases of deliberate deceit is implausible. Reincarnation is a standard tenet of belief in many Eastern religious philosophies; it was therefore commonly met with in the writings of the ancient Greek philosophers, like Plato, Plotinus and Pythagoras. The concept formed part of the philosophy of the Christian theologian, Origen (185-254 CE), and of Christian belief until it was outlawed by one of the several Councils of Constantinople (in 553 CE).

The principle of reflection can also be regarded in the sense suggested by Socrates that 'The unexamined life is not worth living'. We need periods of calm reflection and meditation to bring us back to the balanced centre of holistic living. We need to break the cycle of perpetual motion that modern consumerist economics demands [175]. We can clarify various issues rationally but the input from intuition can also be significant and meaningful. Only in a relaxed meditative state can we become aware of the inner self. We need to believe in ourselves and what we can achieve, and not be reticent about asking for spiritual help if our intentions are honourable. Such help is most effective if we put ourselves in the frame of mind that the help is forthcoming and that we will succeed. All of the Eastern mystical philosophies have meditation at their core, just as prayer is an integral part of Western religion; but where Eastern meditation is a reflection on life, past and future, to help the individual to find the Way, prayer in the West too often is about asking for outside help to solve problems encountered by the supplicant.

The principle of action, which may also be called *the principle of purpose* or *the principle of dhamma*. Whatever we may have debated with ourselves or with others about a certain course of action, ultimately we actually have to get it done. We may not yet know the path, but at least we can set the goal or purpose and begin our journey. We are all built with an enormous potential to achieve – perhaps physical things like playing a sport well or being a builder, perhaps academic achievements contributing to knowledge in the world, perhaps helping others to acquire knowledge through teaching or writing but, whatever it is, dreams, intentions, ambitions alone will not get the job done. Caring for our health, meditating on our successes and failures and resolving to improve on our shortcomings, thinking and planning our days, our whole life, are all admirable provided we then use this raw material to make a difference in the lives of others. Somehow we have to find the internal spiritual energy, what Greek philosopher Aristotle called *entelechy*, to turn our potentiality into actuality, and the universal spiritual field of energy is there for us to draw on to help us to do just that. We should be aware that the universal spiritual energy is infinite. If there are challenges in our way, our satisfaction will be that much greater in overcoming them. Once we have found the way by reflection, the most effective technique is to concentrate on the solutions, rather than focusing on the problems.

Often, the easiest path is one of inertia, letting others take care of the challenges while we go out and enjoy ourselves. But we need to remind ourselves that everything we do – and everything we don't do – has an impact globally. This aspect of living holistically embraces also the *principle of integrity* – doing that which we know intuitively is morally right – and the *principle of compassion* – living for others and deriving our fulfilment from theirs. Another process of dedication to the Spirit would be using environmentally friendly products wherever possible so that we make a contribution to the welfare of planet Earth. Essentially it is trying to live according to the Eightfold Way of the fourth Noble Truth of Buddhism, described below. The principles of the Buddha's teaching are known as *dhamma* or *dharma*. This practical awareness of service to an Infinite Being, whatever form we may envisage that as taking, is enshrined in the tenets of most religions. The same divine spirit dwells in the synagogues of the Jews, the mosques of Islam, the cathedrals and churches of Christendom, the temples of the Hindus, and the sacred sites of spiritualists and theistic pagans. This divine spirit transcends space and time, beliefs and traditions, cultures and customs.

Eastern spiritual philosophy

When we look at these principles of spirituality, we see at once that they are most in harmony with Eastern spiritual philosophy. In the West, the focus of religion is on the individual – most of all in Christianity, least so in Islam. In the East, the focus of spirituality is on the community – most of all in Confucianism, least so in Taoism. The Eastern religions with the greatest number of adherents are Hinduism and Buddhism claiming some 14% and 6% of the world's population, respectively [www.adherents.com, 2005]. Although 16% do not belong to any organized religion, at least half of these still believe in a divine agency and only some 2½% regard themselves as atheists. To underline holistic principles embracing spirituality it is helpful at this point to focus first on the Eastern religious philosophies.

Hinduism

Hinduism is probably the oldest of the world's formal religions, beginning some 2000 BCE [328]. Although Christianity tends to regard Hinduism as a form of atheism, it embraces concepts parallel to the Western ideas of God and soul. The state of *moksha* or *mukti* (enlightened consciousness) is attained by extinguishing worldly desires so as to achieve union of individual soul or self (*atman* or *jiva-atman*) with the Absolute or *Brahman*, as laid down in the Hindu scriptures. The earliest of these, the four sets of *Vedas,* are believed to have been written down around 1200-1000 BCE after centuries of oral tradition. These make up the *Shruti* that are believed to have been transmitted to the Gurus or *Rishis*, the sages who served as channels of divine knowledge (Sanskrit: *veda* = knowledge). In the *Vedas* we are told that we and the world comprise *purusha* (soul, spirit, self; Communal Soul) and *prakriti* (the potential(ity) of Nature, including all matter; the non-self) out of which the world evolves under the influence of *purusha. Prakriti* may be changed from one form to another, but never created or destroyed: this is a spiritual counterpart of what the scientific world would recognize as the First Law of Thermodynamics. The best known of the *Vedas* in the West is the *Rig Veda*, a collection of hymns praising the deities.

The *Upanishads* also form part of this scriptural literature and they are thought to have originated around 400 years BCE: these contain the elements of Hindu mystical philosophy. They were transmitted by the Gurus to their students much as the masters of Greek philosophy instructed their students. The *Brihadaranyaka Upanishad* defines *atman* as a universal spirit or vital force and, as such, operates in every form of life, not just humankind; *Atman* is therefore at times identified with *Brahman* the God of Hinduism, without attributes, unknown and unknowable. Hindus therefore have their *avatars*, *Rama* and *Krishna*, who are manifestations of their god on Earth. Their exploits are described, respectively, in the epic poems, the *Ramayana* and *Mahabharata*, which make up the *Smriti* of Hindu scripture.

The longest poem of the *Mahabharata* is the *Bhagavad Gita*, 'The Song of the Adorable One', and it is the best known of the epics from this literature. The *Bhagavad Gita* says that life should be devoted to selfless work, carried out from a principle of duty and not undertaken for personal gain. The maxim is: 'plain living and high thinking', to be satisfied with what we have and not be continually questing after more of everything. Some branches of the Hindu religions place more emphasis than others on the devotional meditation (*bhakti*) aspect of the faith as a means of enlightenment, but Hindu philosophy is capable of many interpretations. One thing that is certain is that the principles it advocates are the antithesis of the materialism that pervades so much of Western society. There is much more emphasis in Eastern religion on the attainment of the wisdom of the sage. The goal is to strive towards the wisdom of accumulated learning in old age, of making informed impartial judgments, and of recognizing and accepting the uncertainties and limitations of human existence, helping others as much as possible along the way.

Human existence is prone to error or evil through ignorance (*avidya*), so one must seek knowledge (*jnana*) in earthly existence, but also show devotion or faith (*bhakti*), and the practice of useful work and good deeds (*karma*), as described in the *Vedas*. To follow *bhakti* without *jnana* leads to vacuous sentimentality; but *jnana* without *bhakti* is just specious dialectic. To study and adhere to dogmas, to recite creeds and practice rituals as prescribed by any religion is meaningless without the *karma* and the attitude of *ahimsa* (non-violence, non-hatred) to go with it.

The path to *moksha* involves a quest for *anatman*, non-self, through meditation and *yoga* (mental and physical discipline). *Yoga* is a Sanskrit word meaning 'yoking', that is, 'joining two things together', namely, body and spirit. One of the ubiquitous practices of followers of all of the Eastern religious philosophies is meditation. The aim is to dissociate oneself from one's material surroundings into a spiritual plane of tranquillity. At the very least this reduces stress and thereby produces measurable positive benefits to health. Those who become highly skilled at the processes of meditation claim that it produces spiritual refreshment and enlightenment through glimpses of the ineffable. *Yoga* has attracted many western enthusiasts as a means of keeping fit, and because it represents an oasis of calm and an opportunity to escape the stresses of modern life in tranquil surroundings. It enables practitioners to realign their sense of values of what is important in life by making time for a spiritual component.

There are different types of *yoga* practice that focus on different aspects of spirituality. *Hatha yoga* focuses on the familiar *asanas* or postures and exercises to promote energy flow. This is the 'keep-fit' type of *yoga*. *Karma yoga* concentrates on the doing of good works for the benefit of others – a life of selfless devotion to the wellbeing of other people. *Bhakti yoga* is for those who wish to live their lives as a devotion to the ever-present divine; and *jnana yoga* concentrates on acquiring knowledge of ultimate spiritual truths. Both of these are more suited to disciples of the monastic traditions. *Purna yoga* is a form of contemplation advocated particularly by the Indian mystic Sri Aurobindo (1872-1950). It seeks to maximize living practically but harmoniously in the real world, at one with one's environment, without the rigours of extensive study or meditation. Ideally, it would be good if we could embrace all of these, but as each requires almost a lifetime of study and practice to perfect, realistically one must choose one's preferred path to enlightenment.

Achievement of a contemplative or meditative state is assisted for some by tranquil music in the background, or by the chanting of a repetitive phrase, called a *mantra*, or by fixating on a symmetrical geometrical design, usually encompassed by a circle, or occasionally a square, called a *yantra* or *mandala* (Sanskrit: magic circle). The designs symbolize mathematically the universal divine energy and help to focus the meditator's own mind and internal energy. Geometrical harmony is believed to be more conducive to calm meditative states than asymmetric designs: and the same could be said of aural or visual harmony in music or painting. The idea is to focus on the shape for as long as necessary to be able to hold the essential details in the mind with the eyes closed.

> To fully understand, one has to sit down and wait for the universe to enter.
>
> Brian Swimme, *The Hidden Heart of the Cosmos*, 1996

The Swiss psychoanalyst Carl Jung believed that *mandalas* were of great psychological and religious significance because they reflected some deep innate spiritual truth buried within the subconscious. Many of his patients reported visualizing such designs in dreams or trance states, and Jung drew several such designs himself to help him when he was in states of stress. They represent the archetypes transmitted through the collective unconscious, discussed in Chapter 2.

Hinduism can be thought to have certain features in common with Christianity: the concept of an interactive though unknowable God (*Brahman*) and a Godhead (*Brahma*); the Trinity (*Trimurti*) of Brahman (God), Atman (Holy Spirit), and Vishnu, of whom Krishna and Rama are avatars (incarnations or manifestations); the concept of the personal human soul (*atman*); a division of scripture into the *Shruti* (*Vedas*, which include hymns (*cf.* Psalms) and *Upanishads*; *cf.* Old Testament) and the *Smriti* (*Mahabharata* and *Ramayana* – exploits of the avatars; *cf.* New

Testament); supplication to an external God (*Brahman*) or icons similar to images of the Virgin Mary and the saints to whom Catholics make their devotions. Many of the principles of Hinduism described above accord perfectly with holistic living.

Buddhism

Buddhism, which evolved within a society where Hinduism was predominant, originated with Siddharta Gautama (Sanskrit) or, in Pali, Siddhattha Gotama (*ca.*563-*ca.*483 BCE), who was brought up within an aristocratic and noble, possibly royal, family in Northern India [140; 181; 201; 323]. He was born just inside the border of present-day Nepal amongst the Sakya people. Hence he is sometimes referred to as *Sakyamuni* (the sage of the Sakyas). Details of his life and teaching were not fully explored and written down until some 500 years after his death and there are only fragmentary details in the Pali Canon, the basic scripture of Buddhism. The epic poem called the *Buddhacarita* or 'Acts of the Buddha' by Asy̧aghoṣa was not written until the 1st Century CE, by which time some biographical details had been elaborated! Biographies are a relatively recent Western creation. It is almost certain that biographical details of the life of Jesus, compiled in the 60 years or so after his death and presented in the New Testament, were similarly creatively enhanced to convince non-believers of his spirituality. It was not until the 2nd Century CE that images of the Buddha became popular; before then he was represented symbolically by a wheel or a tree. The title 'Buddha' means 'the enlightened one'. The story goes that Gautama was confronted by the sick, enfeebled and the dying at the gates of his palace. He realized that all his wealth could not prevent his experiencing the fate common to all humankind – that of suffering and death. At this point he experienced his first spiritual transformation – to explore the reasons for suffering. Thence he renounced the life of luxury into which he was born in favour of one of poverty and humility and he became a wandering ascetic in his search for the truth of suffering.

Subject to normal bodily needs of food and shelter which intruded on his contemplation, he realized that he had not gleaned any insight into the nature of human existence in either of these extremes which he had experienced — luxury or poverty. It was at this point he underwent his second spiritual transformation resulting in his enlightenment (*bodhi*) (in Chinese: *tun-wu*; in Japanese: *satori*). His teachings are known as the *Dharma* (or *Dhamma*) and were written first in Pali (and hence called the Pali Canon) and then other principles were added in Sanskrit.

Thereafter Gautama began his life of teaching which he continued for the next 45 years and in the course of which he became known as the Buddha, the Enlightened One. He lived just at the time when a distinctive culture was emerging in India, while contemporary teachers were expanding spiritual horizons within civilizations in China, Greece and the Middle East [8]. After his enlightenment, the Buddha renounced the life of extremes and instead advocated a 'middle way' as the path to spiritual fulfillment — neither total abstention nor over-indulgence. He advocated satisfying the body's physical needs to live in comfort so that one can then lay aside considerations of the body to focus on the purification of the self. The Buddha never saw himself as anything other than an ordinary man, certainly not possessed of any divine authority. He preached from his own life experience and encouraged others not to simply accept his teachings but to examine them critically within their own personal context.

Some of the principal features of Buddhism[1] – the concepts of *samsara* (cycle of successive reincarnation) and *karma* (*Sanskr*: action – cause and effect in a Buddhist context; the driving force behind s*amsara*), and the principles of non-violence and the sanctity and interconnectedness of all the natural world – are taken directly from Hindu belief. The word *karma* first appears in the *Rig Veda* where it means sacrificial action to help others. *Karma* is all about intentionality of action – physical, verbal or mental. The kernel of the Buddha's teaching was the *Four Noble Truths* that lead to enlightenment (*nirvana* (Sanskr.) or *nibbana* (Pali));

different texts give different interpretations of these teachings.

The first of these is the truth of *dukkha*: it is an inescapable part of the human condition. *Dukkha* is usually translated as 'suffering', but it can also be rendered by unease, disquiet, discontent, dissatisfaction or imbalance. It is the complement of *sukha* (satisfaction). There are three elements that comprise *dukkha*: (i) actual physical pain; (ii) change or impermanence (*anicca*), represented by Hegel's *Veränderlichkeit,* and (iii) the state of *avidya* (ignorance, confusion, unenlightened existence). Buddhists accept that this is a natural and inescapable part of the human condition. But we must not allow ourselves to become too focused on *dukkha*, for this can lead all too easily to a very negative outlook – the very opposite of our way to fulfillment and holism.

The second of the truths of existence is the truth of *samudaya* (arising), that is, there are specific causes for *avidya* and the *dukkha* that this creates. Overall, the problem with the human condition is our propensity for *kleshas* (afflictive emotions and thoughts) that predispose us to craving (*tanha*). *Tanha* shows itself again in three ways: (i) in a continual questing, a constant yearning for something different, something better, like wealth, power, status, physical gratification from food or sex, which we believe will give us pleasure. This may be considered as comparable to Alfred Adler's psychological image of the human personality as one that strives for superiority or recognition. In a more aggressive vision it is comparable to Friedrich Nietzsche's will-to-power. The objective of our craving is always a desire for pleasure, at some level. (ii) Another source of *avidya* is our yearning for eternal life, and empirical evidence would seem to indicate that there is indeed some form of individual existence after death [99; 387; 395]. (iii) The third aspect of *avidya* is a longing to end our mental anguish or physical pain by non-existence. This is one of the more negative aspects of Buddhist thought – few people have a desire for mortal death unless their lives are filled with extreme mental or physical suffering, and this is not a condition we would expect to meet until we are nearing the natural end to our mortal existence.

The third of the Noble Truths is the truth of *nirodha* (cessation, escape), that is, it is possible to escape from *dukkha* into a state of bliss or *nirvana*. This is achievable through a sublimation of self, a suppression of these natural, human dispositions by a process called *anatta*, in which we focus our being on the Ultimate Reality beyond our selves. The Buddhist belief is that we are happiest in this state of *anatman*, non-self. This might be so for the eastern mentality, or for westerners living a life of seclusion or relative isolation, but for those involved in normal western society, urban or rural, the ultimate goal of individuation, fulfillment of Self, as suggested by Jung seems a more realistic and indeed desirable goal of existence. Our identity, our ego or self is all we have in conscious awareness so to aim for the annihilation of this ego seems to be a denial of life. Buddhist Christmas Humphreys has views that endorse the Jungian position (see below). This again is a rather negative aspect of Buddhist philosophy. What we need holistically is to fulfill Self by service to others.

The fourth of the Buddhist Noble Truths is the truth of *Magga*: that the path of escape from *dukkha* or discontent is through the Eightfold Way of right living in 1. beliefs and understanding; 2. motives and desires; which together comprise wisdom (*prajna*); 3. thoughts and language; 4. deeds or activity; 5. profession or livelihood; which make up ethical activity or morality (*sila*); 6. attitude or awareness; 7. insight or mindfulness; 8. discipline or concentration; which comprise meditation (*samadhi*).

Sometimes the Fourth Noble Truth is called the Middle Way.

1. Right beliefs and understanding means seeing the world as it is and not deceiving ourselves with false notions: having the courage to face reality. It is sometimes presented as 'right view'. This includes not deceiving ourselves with man-made religious fantasies.

2. Right motives or resolve means setting ourselves worthwhile

goals that will truly uplift us and others: subjugating emotion to reason. We can be guided by others but we can only find Truth for ourselves.

3. Right thoughts, expressed in speech, is thinking positively about life situations, and saying good of others, while refraining from lies or malicious and futile gossip.

4. Right deeds is acting in accord with these precepts for the benefit of others and not to their disadvantage; and living primarily for the benefit of others, abjuring selfish goals: gaining fulfillment through enhancing the lives of others; to embrace wherever possible the qualities of *dana* (giving) and *metta* (loving kindness).

5. Right profession gives us the day-to-day framework in which we are able to practise ethical action (*sila*).

6. Right attitudes is a call to probe more deeply into the subconscious, to make strenuous efforts so that right thoughts and deeds become second nature.

7. Right insight (*vipasanna*) or mindfulness is a call to heed the inner voice, the primitive self as well as the developed rational self, to extend the range of conscious awareness of other people and the world around us.

8. Right concentration through discipline means cultivating this insight by introspection and meditation (*samadhi* or *bhavana*), assisted by the use of chants (*mantras*) or fixation on regular geometric designs (*mandalas*); learning to focus on relevant issues to accomplish positive action that benefits ourselves and others.

The objective of the Eightfold Way is to develop wisdom (*prajna*) and compassion (*karuna*) to achieve the state of *nirvana*, free of worldly craving and discontent with an understanding of the true nature of mind. In less spiritual terms we can say that the aim of Buddhism is to produce happiness for its followers. In this, Buddhism presents an admirable path to a spiritually based holistic existence. There are, however, negative aspects of this philosophy as mentioned above and, as with other spiritual outlooks, there are insights we might wish to endorse and follow as far as possible and others that may not seem so acceptable or appealing. The feeling of being connected to a spiritual realm, be it deity or universal spiritual energy, is an important contributor to happiness, which we achieve by social relationships on the physical plane. Such connections boost the feeling of self-worth which in turn adds to contentment. We might recall the words of H.T. Hamblin quoted in the opening of this chapter about well-being arising from service to others. We have become too self-centred, too anthropocentric without looking for our place in the natural world.

The present Dalai Lama believes that there are strong similarities between concepts in Buddhism and those of quantum physics, as suggested by Capra [44] and Zukav [403]. Both Buddhism and science put empirical experience first, to be interpreted by reason [171, p.25]. There are many different world-views described by the scriptures of the various religions, just as there are diverse theories of the natural world in science. Although in western religion one is expected to adhere without question to the principles of scripture, in Buddhism scripture is taken to present guidelines to a world-view just as scientific theories are to the experimentalist. Each *bodhisattva* must find his own Way, just as each scientist does. Meditation serves the same function in Buddhism as introspection in science: both involve faith – in the Eightfold Way or the validity of the scientific method – and both involve communion of the soul with the universal spiritual field. Inspiration is as important to the scientist as intuition to the seer. As I remarked previously [193, p.280], we have no grounds for disbelieving seers and prophets concerning the existence of the spiritual realm just as we must accept the word of scientists and mathematicians regarding their discoveries, unless we are experts in the field [171, p.30]. Where science distinguishes between life and non-life on the basis of entropy (living organisms are those that increase order or decrease entropy in their structure, namely all plants and animals), Buddhism distinguishes between sentient and non-sentient beings (animals, including humans, on the one hand; plants and rocks on the other). In a holistic world-view, everything is made of the same atoms – animals, plants and rocks – and it is only in how they cohere and interact with one another (as atoms and as beings!) that they differ.

The belief in some form of spiritual creative force or energy is one that is held by nearly 90% of humankind. Such an energy, in the form of the zero point field or quantum vacuum, is suggested by rational investigation of the natural world; it is a concept that is also endorsed by most psychics, seers and sages. As stated above, it is generally held that in Buddhism, there is no concept of soul or God – only a Universal Mind as guardian of the *karma* for the next incarnation, though such a universal energy could also be interpreted (though a Buddhist might not do so!) as an all-powerful creative agent. So we should qualify this position with a quotation from The Buddhist Society's Twelve Principles of Buddhism which states: 'Buddhism does not deny the existence of God or soul, though it places its own meaning on those terms'.

The God of the Buddhist is that principle from which the universe was born. Soul is that characteristic of an individual that progresses towards enlightenment. In his books on Buddhism [179-181] Christmas Humphreys Q.C. (1901-1983), a High Court judge and founder of the Buddhist Society of Great Britain takes issue with the concept of no-self or no-soul, which he considers a negative and dismal teaching and says was never taught by the Buddha [179, p.42]. He argues that there must be a 'self' or ego, a personal identity as envisaged by Freud and Jung, to undergo enlightenment and implement the *karma* from past lives. As fundamental in this process, Buddhism encourages a mind-set that eschews greed, hatred and delusion, the Three Fires that lead to *dukkha*. Failure to achieve this state in one lifetime encourages the enactment of *karma* (or *kamma* in Pali) to learn these lessons in subsequent lifetimes. But *karma* is not a punishment, and all of life's misfortunes should be regarded as part of the individual's learning process. Remember

Pradervand's comment given earlier about trials being blessings in disguise.

Buddhism, despite having no belief in a creator God (since the universe is eternal and infinite) or personal soul in the orthodox philosophy, is close in many respects to the one-world vision of paganism and holism; Buddhism does accept the existence of gods and spirits and even in some interpretations (see above) a personal soul. In a theistic world-view we can take from Hinduism or western theology the idea of a divine agency acting as Creator and Designer of the universe, the spirit of which permeates the whole of creation – which spiritualists and theosophists will designate as the universal spiritual field or akashic field and secularists may regard as the zero point energy field or quantum vacuum. If we accept Christmas Humphreys' interpretation of Buddhism and his assertion of soul and take on the personal responsibility for enlightenment advocated by Taoism (see below), the Eightfold Way presents an admirable spiritual path to holistic living.

Taoism

Taoism and Zen are derivatives of Buddhism that arose in China and Japan, respectively, though the indigenous religion of Japan is Shinto ('the Way of the gods'). The gods, of which there are literally millions, are *kami* or *shin*; *to* is the Japanese version of the Chinese *tao*: Shinto has some ten million adherents. Thus *Shinto*, like the *tao*, represents a path to personal enlightenment and fulfillment, a holistic programme for living, as with other religions or spiritual philosophies. The *Tao* is also the spiritual unity and creative energy of the cosmos. The Buddha's *dharma* teaches pupils to follow the middle way, just as Jesus is reported as saying 'I am the Way'. The word 'Way' or *tao* carries with it an intrinsic spirituality.

The name of Lao Tzu is usually associated with Taoism because he is regarded as the author of the *Tao Te Ching*, though it is by no means certain that such a historical person ever existed. If he did, he would have flourished in the sixth century BCE and could possibly have been a tutor of Confucius. However, there were a number of advocates of this philosophy living in the same period: Yang Chu (440-360 BCE) and Chang Tzu (399-295 BCE) also perpetuated these ideas and their lives are better documented. It is possible that the origins of Taoism may go back as far as those of Hinduism – to some 2000 years BCE.

The word *Tao* means 'way' or 'path'; *Te* is the creative intuition of humankind that produces great works of art or spiritual insight of the unseen dimension of existence; *Ching* is a book or text. So the *Tao Te Ching* is a text that expresses the wisdom of a seer or school that points the way to spiritual enlightenment.

> The *Tao* that can be talked about is not the Eternal *Tao*.
> The name that can be named is not the universal name.
>
> As the origin of heaven and earth, it is nameless.
> In the infancy of the universe there were no names.
> Naming fragments the mysteries of life
> into ten thousand things and their manifestations.
>
> Yet mysteries and manifestations spring from the same source:
> The Great Integrity which is the mystery within manifestation,
> the manifestation within mystery;
> the naming of the unnamed and the un-naming of the named.
>
> When we become aware of these interpretations,
> we will pass through the gates of naming notions
> in our journey towards transcendence.
> Lao Tzu, *Tao Te Ching*, Verse 1: Transcendence

We are reminded that the world is made of the manifest and the hidden – that which we can perceive with our senses and reason, and that which we can know only by intuition in quiet repose and meditation to let our soul find expression.

Taoism sees the world as comprising the two complementary spiritual forces: the feminine *yin* (representing softness, suppleness, intuition, sensitivity, light, heaven, the spiritual) and the masculine *yang* (representing the assertive, rational, aggressive self; Earth, the material). The aim is to live quietly and lightly, to intrude on others and the environment as little as possible, in balance between *yin* and *yang* and in harmony with the natural world, and to encourage others to live thus by example: to overcome evil by persuasion not aggression.

Only simple and quiet words will ripen of themselves.
For a whirlwind does not last a whole morning,
Nor does a sudden shower last a whole day.
Lao Tzu, *Tao Te Ching*, 23: Sincerity

The *Tao*, within its many gems of wisdom, contains an exhortation that is so appropriate and so urgent for us today – to focus less on our materialistic fragmented lives and look within and beyond to the spiritual whole.

The Great Integrity is infinite, undefined and silent in its Primal Simplicity.

Nothing is its superior, and Earth and Spirit will be harmonized
in the sweet daily dew when humankind embraces It.

When the Primal Simplicity fragmented into myriad pieces
With their myriad names
Earth became endangered.

Are there not enough fragments now?
Is it not time to stop and for all to return to the universal ocean
from which all streams once emerged?
The *Tao* is to return to the Great Integrity so that

we no longer need our list of endangered species,
and Man and Earth can live again in harmony.
Lao Tzu, *Tao Te Ching*, 32: Unity

The *Tao* is the path or Way of spirituality that leads to enlightenment – how to live holistically in the world and persuade others to do likewise by example and wisdom teaching, but never by commandment or claimed authority, and always with respect for other people, with their customs and beliefs, and for the rest of the created world. We must acknowledge that we need both the material and the spiritual in our lives to find happiness and holistic fulfillment. Whatever knowledge, whatever benefits science can bring to us by its reductionist approach, ultimately we must put the fragments back together again and see the undivided whole.

Taoism directs its teaching to disciples to learn by example from what they see and hear in the external world around them: but they must find the Way for themselves.

He who knows others is wise;
He who knows himself is enlightened.

He who conquers others has brute force;
He who conquers himself has inner strength.

He who knows when he has enough is wealthy;
He who follows the Tao diligently is wise.

He who finds his true purpose will live long;
He who lives a truly creative life will leave an eternal legacy.
Lao Tzu, *Tao Te Ching*, 33: Identity

Knowledge of the world is important, for we need this to live successfully with others; but the ultimate wisdom is understanding of oneself and our true purpose in life to care for all around us. As the ancient Zen homily says: 'Do not follow in the footsteps of the Masters: seek what they sought'.

The *Te* may also be interpreted as the fundamental force or energy of the universe, present in all things as *chi.* This could be compared conceptually with the Christian idea of a creative Godhead manifest as Holy Spirit; or the emanations of the *sephiroth* from *En Sof* in Judaism; *Atman* the breath of *Brahman* in Hinduism. But the *Te* is not in any sense a god but a secular creative spiritual life force, like the zero point field of quantum physics. It is the only ultimate reality; for how can the world be real if the Self that is observing it is intangible?

Confucianism

Confucianism arose in China at around the same time as Buddhism in India. Its founder, Kung-Fu Tzu or Confucius, lived from *ca.* 551 to *ca.* 479 BCE during the Chou dynasty (*ca.* 1100 – 256 BCE). Most of the information we have about his life came originally from a biography that made up just a single chapter of a book *Shih chi* (*Records of the Historian*) that was written, again, as with the Buddha, long after his life, in the 1st century BCE by Ssu-ma Ch'ien. The Chinese scholar Ts'ui Shu (1740-1816) considered much of this work unreliable so wrote a biography of his own which has become the standard source of reference on Confucius. The sacred writings in Confucianism are called The Classics comprising a set of six books that are thought to have existed in the time of Confucius and on which he may have actually worked himself. The six books covering the history of other Chinese sages besides Confucius are called the *Songs*, an anthology of poems; the *History*, a set of official government documents; the *Rites*, giving details of ancient Chinese rituals; the *Changes*, a manual of divination; the *Spring and Autumn Annals*, a chronicle of the feudal state of Lu where Kung-Fu Tzu lived; and the *Music*, which has been lost. As with Buddhist sacred literature, in the millennium after Confucius, these books were added to; the book of sayings attributed to Confucius, the *Analects*, comes from this period.

The philosophy of Confucianism centres on uplift of the individual through the education of society in order to instill a sense of moral values and mutual responsibilities within the family and in society and government. As might be expected, the Golden Rule finds expression in the philosophy of Confucius. When a pupil asked the Master for one rule to follow as a conduct through life, the Master replied: 'Do not do to others what you would not want done to yourself' [*Analects*, XV.24]. Confucianism encourages respect for the experience and wisdom that comes with age and authority and so lacking in western society.

* * *

Although the Eastern religious philosophies of Hinduism, Buddhism, Taoism and Confucianism are quite different in detail, the central theme is enlightenment of the individual through introspection rather than through instruction by authority, though a thorough knowledge of the sacred texts, like the Confucian *Classics* or the Buddhist *Pali Canon,* is fundamental for all adherents. They all stress the need for societal coherence and for mutual benevolence, they respect all living things, and all have great concern for the natural environment in a holistic existence. Humankind and Nature are regarded as being at one with the Universal Mind or Spirit, the akashic field that holds the *karma* of every individual.

Overall then, in its view of life as a struggle and its encouragement of unselfish existence, and its plan for charting a path to such existence by careful examination of our lives through rationalization and meditative insight, Buddhism enumerates many features of our lives to which we would do well to pay attention for holistic living. A vision of the oneness of the universe and our lives as part of the spiritual continuum of Nature is an essential component of holism. Combining this practical philosophy of everyday life laid out by Buddhism, with that of concepts of the divine and individual soul from Hinduism, individual responsibility and example from Taoism, familial duty, right education and social involvement from Confucianism, and environmental concern from all Eastern philosophy, we have as near a perfect paradigm for holistic living as we can probably get.

The mystical dimension of western religion

The major western religions all have their mystical sects, but these are smaller movements that have developed away from the mainstream of the faith. In the West, God is regarded as so utterly unattainable to the ordinary worshipper that it became necessary to create intermediaries – priests, the pope, rabbis, imams – who, it was considered, were closer to an understanding of God's will. Christianity took this concept one stage further by creating an image of God's incarnation on Earth in Jesus. But this vision of the inaccessibility of the Divine to ordinary people also tended to isolate people as individuals and focused attention away from the coherence of society and the unity of humankind with all of creation. The mystical faiths broke down this barrier of insulation between the individual and the divine. Western religion has done humankind a disservice in focusing on the relationship between Man and God, while ignoring the relationship of humankind with the spiritual universe. Indeed, any suggestion of spirituality outside of the human domain is anathema to orthodox western religion. The techniques of scientific reductionism too have rejected this spiritual holism, but it is quite possible to be both rational and mystical [142, p.21;193, p.224].

Whatever one's religious affiliation, a more mystical approach to the Divine or to the spirituality of the natural world is an essential component of a holistic lifestyle. Consciousness is the ground of all Being and is the only image of God which has any rational coherence. There is no trial and retribution after earthly existence, no punishment or rewards, just pure spiritual Being.

The symbolism of the Kabbalah has a central role in contemporary spiritualism, with its focus on the Tree of Life, the unity of the created world, and the transcendence of the spiritual. Gnosticism is now of essentially historical interest only, though it did have an influence on Carl Jung. But it is important because Gnostic texts provide some different insights into people and situations described in the Bible, which many use unquestioningly as the basis of their spirituality in all western religions. The view presented of Jesus in the Gnostic texts is much more like that of the Buddha, or that described by Emanuel Swedenborg – an inspirational leader to serve as a guide to spiritual understanding and wisdom, to dismiss illusion and replace it with enlightenment. Swedenborg's ideas of the Second Coming as a realization of the ideals personified by Jesus [355] are highly relevant in the context of Gnosticism. The Gnostic texts present a more holistic view of the Divine that is much less remote than the God of western religion.

Sufism shows how immersion in one-world spirituality is not incompatible with living meaningfully in the material world, especially when the case is presented as eloquently as it is by H.J. Witteveen [see 397, 398]. This is particularly important in view of the violent extremism currently associated with Islam. Even if the Koran is not interpreted in this light by the majority of Muslims, in the eyes of adherents of other faiths, Islam has become associated with *jihad*, and the Cold War against Communism has been replaced by ongoing violent confrontation with Islam. The Prophet was a man of peace and tolerance of other faiths based on the words of the earlier prophets. He even married a Jewish widow, Rayhanah bint Zayd as one of his wives The concept of *jihad* has been misappropriated by extremists: it is intended to mean 'struggle' not 'military action'. The struggle is both internal in living the righteous life, and external in its opposition to cultures like that of some countries in the West with their materialism, sexual promiscuity, drug culture and gun law.

Devotees of Kabbalah, Gnosticism and Sufism all focus on direct experience of the Divine or the universal spiritual energy or the ultimate reality behind the material world of the senses, however we want to express it, and allowing this experience to influence the behaviour of the adherent for the benefit of humankind, both contemporaneously and for those living in the future. These divisions of Western religion share with the mainstream of Eastern religious philosophy the approach to direct communion with the Divine Spirit; perhaps, in this, they present the best hope of convergence between the various world faiths. Such mysticism may involve awareness of a being that the percipient regards as divine or a manifestation of the Divine (in the West), or may simply involve transportation of the mind (or soul) to a dimension of being that is distinguished as ultimate reality (in the East). This facility is available fully to only the few whom we call prophets, seers, visionaries, or shamans. Most of us however can achieve this to a limited extent by meditational practices or by spiritual transport through aesthetic pleasure. For all of us, the universal spiritual energy is more readily accessible through our own artistic creativity, or an appreciation of that of others, or immersion in what we take to be the created natural world. Educationalists have come to realize that having children participate in music making, art or crafts is an essential part of their development, even if it seems to be of little 'use' to society.

Kabbalah

In the Jewish faith there is a mystical movement represented by the Kabbalah (also Cabbala or Qabbalah), which means 'received tradition'. Originally, it was handed down orally from one generation to the next in the centuries BCE with its roots in the books of the Bible, but from the 11th century there has been an expanded written tradition originating in the scholastic milieu of al-Andalus[2]. Although the region was under the control of an Islamic administration, there was an exemplary toleration of all western religions and cooperation between the faiths in a spirit of furthering scholarship, particularly in trying to revive the learning of ancient Greece and Rome that the Christian Church had done its best to extinguish [162; 252].

Kabbalah sees the world as one percent material, accessible to the senses, and ninety-nine percent hidden, which is the realm of spiritual light[3]. This spiritual domain is the realm of Plato's Forms, of Hegel's *Geist*, of Jung's collective unconscious or of Sheldrake's morphic field – the Communal Soul of the living and the earthly departed with whom mediums communicate [193, Chap.7].

Spiritual healers, spiritualists in general and shamans of indigenous peoples call upon guardian spirits or angels from this spiritual domain to help guide them in their daily lives. They have an awareness that these spiritual energies or beings are with them constantly. In its expression in spiritualism and theosophy, the symbolism of Kabbalah and its representations of the communion between the material and the spiritual planes provide an important spiritual component of holism.

Gnosticism

In general, mysticism in Christianity has been associated with specific individuals rather than with philosophical movements, like the German Abbess Hildegard of Bingen (1098-1179), the German theologians Meister Eckhart (*ca.*1260-*ca.*1328) and Jacob Böhme (1575-1624), the Carmelite nun Saint Teresa of Avila (1515-1582) and her companion in the stricter discalced Carmelite Order, Saint John of the Cross (1542-1591).

Hildegard of Bingen is one of the best known of Christian mystics to people generally in recent years because of the popularity of the spiritual music and poetry she created. Hildegard and the convent at Rupertsberg near Bingen of which she was abbess were famous amongst the most powerful rulers of her day. She was a true visionary: 'These visions which I saw, I beheld neither in sleep, nor in dream, nor in madness, nor with the eyes of the body . . . but wakeful, alert, with the eyes of the spirit and with inward ears' [32, p.429]. She recorded her prophesies and visions in a trilogy that she compiled over the course of thirty years. She also wrote treatises on natural history and one on medicine, *The Book of Simple*

Medicine. She was an accomplished artist and did the illustrations for her own books. Today, her music transports us to that mystical realm she once occupied on Earth.

Eckhart's concept of the *Seelenfünklein* (divine spark of the soul) as a spiritual aspect of the Divine that resides within each of us as soul can be interpreted in a secular context as the universal spirit inherent in us all that survives mortal death. Böhme envisioned the evolving natural world as created by God in attaining to a higher state of perfection. Both of these holistic concepts should make us much more respectful of the natural world, rather than exploiting it heedlessly for our own material ends. It is a part of us, and we, of it.

Saint Teresa wrote her autobiography as the *Way of Perfection*, and her book *Interior Castle* describes that mystical individual retreat of meditative trance as a place of delight, for 'you can enter it and walk about in it at any time, without asking leave from your superiors' – a wonderful metaphor for the spiritual transport associated with meditation. Saint John wrote of the *Dark Night of the Soul* as that part of the soul's journey after death during which it is purified by self-reflection and by God; but the expression is often taken more practically in a holistic sense as the sometimes anguished self-analysis we have to do at critical times in our lives before we make difficult life-affirming decisions of a new spiritual (or practical) path that we must follow. Circumstances in life often force us to take stock of what we have done with our lives and what we want to do with the remainder; sometimes this involves making emotionally difficult decisions that some describe as 'the dark night of the soul'.

Although Christianity has thus had its share of individual mystics through the ages, the most significant mystical movement that survived for a few centuries was Gnosticism. Apart from its influence on Jung and Swedenborg, its principles have inspired others to the present day. The word is derived from the Greek word *gnosis* meaning 'knowledge' or 'insight' and refers to knowledge of ultimate reality as attained by seers, shamans and mystics. Gnosticism arose in the Middle East in the 2nd century CE and is associated with the names of Marcion, Basilides and Valentinus. The core doctrine is the distinction made between the wrathful, vengeful God of the Old Testament and the material world He created on the one hand, and the loving, forgiving God of the New Testament, exemplified by Jesus, who embodies the objective of pure spirituality on the other. The world is viewed as the creation of a 'half-god' or demiurge who exerted malevolent powers through agents described as 'archons'; it is they who keep the soul trapped in the physical body with its propensity for evil. The only escape for humankind was to join a select group who had access to secret knowledge that could bring about redemption by meditation on what it means to be human, and on our cosmic origins and destiny. In principle, this is so close to the fundamental concept of Buddhism of setting aside the material dimension of the world to focus on the spiritual.

Carl Jung was sympathetic to this meditation on the human character aspect of Gnosticism, which he studied intensively for much of the 1920s[4]. The transition of the focus of the mind from the material world to the spiritual was, to Jung, the psyche reaching out to achieve individuation (see Chapter 2), that is, striving for holism for the individual. According to Professor Gilles Quispel of the University of Utrecht, a distinguished historian of religion, 'gnosticism originated in a potentially universal 'experience of the self' projected into religious mythology' [quoted in 280, p.29]. This may be why Carl Jung felt such affinity with the principles of Gnosticism.

Gnosticism shares with Buddhism the view of the human condition as one of suffering and disquiet from which it can be freed only by dissociation from material things to focus on spirituality. To the Gnostics, this could be achieved only through personal experience and direct revelation to each individual. This need, this propensity of humankind to escape into a spiritual haven from the material realm that confronts us each day may be seen as an example of what Greek philosopher Heraclitus called *enantiodromia*, the movement of any dynamic system into an ontologically opposite state. [Philosophers often use the original Greek or German terms so as to avoid misunderstandings in providing an inexact translation.] It is a fundamental principle of science that matter and energy, particles and waves participate continuously in a similar cosmic dance. A holistic world-view embraces the spiritual or mystical alongside the rational in our quest for fulfilment.

The view that is presented in the Gnostic texts of the character of Jesus forms one of the cornerstones of faith in the Church of the New Jerusalem, founded by the followers of Emanuel Swedenborg (1688-1772). Swedenborg was a Swedish scientist and engineer who turned to theology in the latter half of his life after undergoing a mystical experience [25; 355]. The purpose of the life of Jesus, said Swedenborg and the Gnostic gospels, was to serve as a guide to spiritual understanding and wisdom, to dismiss illusion and replace it with enlightenment. Swedenborg maintained that the Second Coming is not about Jesus physically reappearing on Earth but rather it represents the time when there is universal recognition of the true message of the life of Jesus – or that of any of the major prophets of mystical religious philosophy – that we must live in harmony with one another and with our environment. The current increasingly widespread sense of spirituality may represent the kind of transformation of global consciousness that Swedenborg had envisioned.

The images of sin, suffering, retribution, punishment and repentance are fictions created by the power brokers of the Christian Church, which is why they make so much of the Passion of Jesus on the cross. These were not concepts central to Gnosticism. While a certain amount of suffering and distress is an inevitable part of life that we all have to accept, the New Testament attributes this to our lives of sin [Romans 3. 23]. This is the very antithesis of what the life of Jesus stood for. The Gnostics, like many ancient Greek philosophers, Buddhists and other spiritualists say that fear, ignorance, illusion, and ceaseless questing are the sources of our disquiet, not 'sin'. Gnosticism, Jungian psychotherapy, and holism advocate a search for individuation through the insight of self-knowledge and understanding of our true purpose in life: to provide fulfilment for ourselves through helping others who are living now and will be in the future. Enlightenment is not to be found externally in the dogma and rituals of orthodox religion but within. The *Gospel of Thomas* presents the following wisdom, attributed to Jesus: 'If you bring forth what is within you, what you bring forth will save you. If you do not bring forth what is within you, what you do not bring forth will destroy you' [*Gospel of Thomas* 45. 30-33] and 'Recognise what is before your eyes, and what is hidden will be revealed to you' [*ibid.*, 33.11; both quoted in ref. 280, p.135]. This enlightenment through contemplation is the important message of Gnosticism to those seeking a holistic life-style.

Sufism and medieval Islam

Sufism (Sufiism) or *Tasawwuf* is the Islamic mystical movement. Sufis adhere to all the fundamental principles of the faith[5]. They took their name it seems from their wearing of woollen garments (Arabic: *suf* – coarse wool). The practice was viewed with suspicion by the orthodox faith whose adherents preferred garments of cotton, like those worn by the Prophet. Furthermore, the Christian monks wore similar woollen garments, and the early Christian literature, as we have seen above, also made much of gnosis or direct knowledge of the Divine, so this associated Sufis with heretical Christian practices in the eyes of the orthodox Islamist.

The Sufis proclaimed as the cornerstone of their belief the importance of direct intuitive mystical experience of God (*Ma'rifa*) and of absolute reality (*haqiqa*). To achieve this it is necessary to practice contemplation (*mushahadat*) and learning to be receptive to one's intuition (*yaqin*). As there are no words in our human languages that can adequately communicate to others such experiences, they can only be described by allegory and myth, as demonstrated by the poetry of Jalal ud-Din Rumi (1207-1273). Rumi was a famous Sufi mystic who encouraged the inducement of trances by dancing; this involved spinning oneself around to musical accompaniment, so he is regarded as the creator of the 'whirling Dervishes'. Music and dancing were often used to induce meditative states by Sufis, again in contrast to practices of the orthodox. They also attach great importance to the interpretation of dreams. This very attempt at removal of the gulf that separates Man from God, even symbolically, was another reason why Sufism was regarded by the orthodox Islamists as heretical. However, others found in Sufism the opportunity for the closest possible communion with the Divine and a reassertion of the principles of the faith.

Some orthodox Jews and Islamists view asceticism with a certain amount of suspicion because it implies that God was in error in providing worldly pleasures. But asceticism – of which the wearing of wool garments was just one facet – was the essence of early Sufism. As with Eastern philosophy it was considered essential not to become too attached to worldly things. Ali, leader of the faithful from 656 to 661 CE[6], is reputed to have said:

> Asceticism is not that you should not own anything, but that nothing should own you.

The rise of Sufism coincided with a burgeoning of philosophical tests that attempted to reconcile the views of the Koran with those of the Greek philosophers[7].

Mystical Islam may seem rather remote from everyday life, but the Dutch economist, Dr. H.J. Witteveen, a former managing director of the International Monetary Fund, has shown that it is possible to use Sufist principles to live holistically [397, 398]. His books, and especially the later one, show how to live both spiritually and practically to produce, as he puts it, 'achievement, inspiration and integrity in a tough world'.

* * *

Thus, devotees of Kabbalah, Gnosticism and Sufism all focus on direct experience of the Divine. These divisions of Western religion share with the mainstream of Eastern religious philosophy this approach to communion with the cosmic spirit, an essential component of holism. Such a world-view presents perhaps the best hope of convergence between the various world faiths.

Summary

The ideal of holistic spirituality is to be found by taking, from any or all of the great religious systems of the world, the best of the principles that most accord with our own world-view, and letting those ideals shape our ideas. It is not holistic to blindly follow the dictates and dogma of any one religion to the exclusion of all others. We are all too aware of the turmoil in the world at the present time caused by such intransigence. We have seen above that individual spirituality and a sense of awe and wonder at the grandeur of the cosmic unity is at the heart of a truly holistic life-style, whether we follow one of the eastern religious philosophies or the mystical dimension of western religions. The great prophets and philosophers of the past can guide us, but we must look deep within ourselves to find the true meaning of our lives. We live in accord with the principles of spirituality that are as much an intrinsic part of the natural world as the laws of physics.

In the next chapter we shall see something of the part played by myths and traditions as expressions of the human psyche and thereby producing social cohesion of nations or religious communities.

CHAPTER 4

MYTHS AND TRADITIONS

The stories

Tradition is the handing down of stories, myths, rituals and beliefs from one generation to the next. It links each generation with the past, with their history, and with the ancestors who played a part in shaping the current society as it is, and this sense of continuity with the past is an essential component of a holistic existence. It also establishes a bond between the material world of the present and the spiritual world of the past and, for believers, the world of the divine. It is at least as important today as it ever has been that we should be aware of our history; as the American ecological theologian, Thomas Berry says: 'We need to understand where we are and how we got here' [27, p.ix].

The word 'myth' comes from the Greek word *mythos* meaning, as it does in English, an ancient imaginary story with symbolic meaning for a group of people. It is closely related to the terms 'mystery' (Greek: *mysterion*) and 'mystical', from the Greek *mystikos*, an initiate. The myths and traditions are what bind a social group, whether it is a band of religious adherents or a national or tribal group. We saw in Chapter 2 how junk foods have become part of the social traditions of contemporary society. Tradition is important whether it is viewed religiously or secularly. Thomas Berry believes the underlying problem in society today is that we have rejected the myths of scripture as untenable or irrelevant – what Berry calls the 'old story' – without having defined and absorbed a new world-view, a 'new story', of universal spirituality and wholeness to take its place.

There are some myths that are universal, like the tales of Cinderella, the downtrodden and exploited servant, or Faust, an appropriate symbol of our times as a man who seeks power or pleasure at the expense of his soul. There are other tales that are strongly representative of a national culture, like the *Mabinogion* of Wales and the *Kalevala* of Finland. These are the kinds of myths that Kenneth Rees, in his discussion of myths in paganism, calls 'substantive' [309], as opposed to the myths of scripture 'that direct ways of behaving'. The word 'mabinogion' is derived from the Welsh word *mab*, meaning son, and it is translated as 'tales of a boy hero' or simply 'the bard's tales'. The *Mabinogion* is a collection of eleven tales that were transmitted orally through the ages until they were preserved in written form in medieval times and in the native tongue as the White Book of Rhydderch (1300-1325) and the Red Book of Hergest (1375-1425). They remained little known, even in Wales, until they were translated into English in the mid-19th century by Lady Charlotte Guest. Like many myths, they are allegorical tales of good and evil, of love, loyalty, trust, betrayal, and heroism.

The *Kalevala* tales were compiled in the 19th century by Elias Lönnrot (1804-1884), who was a district health officer in the eastern region of Finland. They were a great source of inspiration to the Finnish composer Jean Sibelius. The set of thirty-two poems were published in their first edition in 1835 from poems spoken or sung to Lönnrot or one of his contemporaries during the field trips they made to rural areas over the course of fifteen years. A second edition of the *Kalevala* was published in 1849 containing fifty poems. The rhythm or metre of the poems is trochaic, as in Longfellow's poem *Hiawatha,* which is also a tale of a native people. There is much internal rhyme and alliteration in the poetry, similar to that found in Welsh poetry where it is called *cynghanedd.* The word 'Kalevala' means 'land of Kaleva' or 'land of heroes'. The suffix '-la' indicates place in Finnish. In Welsh, place is indicated by the suffix '-fa': for example, a doctor is *meddyg* and their surgery is *meddygfa.* There are other similar linguistic developments in different cultures that are not directly related to one another suggesting some deeper underlying spiritual connection.

Other secular myths with strong national connections are the tales of the journeying of Odysseus after the fall of Troy, presented by Homer in the Greek epic poem, the *Odyssey*, and the Norse and Germanic people have their *Niebelungenlied* that inspired the creation of *The Ring* tetralogy from Richard Wagner.

Myths are often associated with religions and as such have important cohesive social value within a nation or culture. Some of the great scriptures – the Hebrew *Torah* and the Hindu *Vedas* and *Upanishads*, for example – are written accounts of myths and traditions handed down orally for very many generations and based on historical people, places and events. Other scriptures, like the *Analects* of Confucius, the *Koran* of Mohammed, or the words of the prophet Moroni delivered to Joseph Smith as *The Book of Mormon*, present the insights of a particular sage or prophet as to how we should live our lives as individuals and societies. The stories of the heroic exploits of the *avatars* Krishna and Rama portrayed in the *Mahabharata* and *Ramayana* are seminal works in the Hindu religion. The Bible has its heroic tales in the exploits of David against Goliath [1 Samuel 17] and Samson against the Philistines [Judges 15].

Myth, scripture and religion may be human in origin but they constitute an essential and ongoing part of the cultural development of humankind and cohesion of tribes and nations. Myth has its source in humankind's creative imagination inspired by spiritual insight and intuition. It presents moral tales and encourages a world-view beyond that of sensory experience: this is one of the distinguishing features of humankind. Myths and folk tales are not created to provide us just with an escape from reality; they represent humankind's attempt to live our lives more fully and give us new insights into the meaning of our earthly existence. In our materialistic existence we have lost sight of this spiritual dimension of holistic living: the material and the spiritual are one, as it is still for native peoples.

We cannot live fulfilling lives only on the material plane and myths transport us to other times and places, often to confront creatures that represent ideas of beauty, fear, evil, loyalty, betrayal or other expressions of human emotions. Fairy tales and nursery rhymes are traditional myths that we tell to children to fire their imaginations and often to teach a moral message. The recent popularity of such fantasy tales as *The Lord of the Rings*, the *Tales of Narnia* and *Harry Potter* testify to the excitement children feel for such stories. As Sue Palmer points out, 'they're ideal for introducing children to the rhythms of language and tuning their ears to language sounds' [285, p.114], as well as their function in stimulating the imagination and teaching moral principles. As nursery rhymes are usually sung, she stresses the 'huge importance of music and song in developing children's social and communication skills' [*ibid.*]. The reciting or singing of myths to children by parents or the elders of a tribe perpetuates their traditions and brings children into the social context of the whole community. The chanting of psalms has a similar function in the Jewish or Christian religions.

The history

History and myth are inextricably interwoven. It is history that has brought us to where we are today, and myths and traditions perpetuate the history of a social group and bind its members together. Archaeologists tell us that even the very first civilizations had some concept of a spiritual domain. Bodies were buried with tools, utensils and ornaments for use in some spiritual continuity of life after mortal death that they assumed would be similar to that which they lived on Earth. The more affluent in Egyptian society, for example, were even buried with small human figurines called shabtis made of faience to act as servants in this spiritual life, and they wrote letters to friends and relatives who were no longer with them on Earth [104; 115]. Everything on Earth was only a mirror of the perfect realm of the gods, just as the Greek philosopher Plato (*ca.*428-347 BCE) saw our human concepts as images of the Ideas or Forms within the spiritual world. The ancient Greeks had their special places, *psychomanteae*, where they communed with those who had left mortal life. The belief was that there was another plane of existence where the souls of the earthly departed dwelt.

Primitive people saw themselves and the natural world around them as much more intimately associated with the spiritual domain than even the most religious of people tend to do today. Rock paintings found in the Northern Territory of Australia and engravings in South Australia imply that aborigines were telling stories to one another as long ago as 40,000 BCE about the Dreamtime, *Alcheringa*, when their ancestors lived. The stories, often portrayed in art-works, are timeless in that they perpetuate what Jung called the archetypes – the inborn ways of thinking about the world. As I said previously, the idea of a sky-god whose son is incarnated on Earth and possesses extraordinary powers is far from unique to Christianity. We shall see later how various trees and stone monuments have been associated since pagan times with myths and traditions.

In the fast-moving urban civilization of the West, there is little time for reflection or for dwelling on history. But, increasingly, people are feeling the need to commune with the past. Archaeology has been given a new impetus in excavations that are shown in television programmes. Ancient non-invasive healing treatments like spiritual healing and the use of herbal medicines are once again commonly found alongside orthodox medicine, and pagan images of the Divine as an all-embracing spirit are becoming more popular to replace the excrescences of man-made formal religion. Through such traditions, distinctive for each tribe or civilization, peoples have built up a holistic view of human life in harmony with Nature. Yet, for all their variety, traditions show many features in common perpetuated by Jung's archetypes though the collective unconscious.

The cultural evolution of humankind has been a remarkable process. Creatures that anthropologists recognize as having some human characteristics have been on Earth for about five million years (m.y.). This is just a tiny fraction (about one-thousandth) of the age of the solid Earth, which has been a very long time indeed in the making, and human history and prehistory represent just a tiny portion of this time scale. Our own species, *Homo sapiens*, of which Neanderthal man is one of the earliest European examples, appeared towards the end of the last interglacial period, and the beginning of the last great Ice Age, some 100,000 years ago. During the last of a series of Earth glaciations, the ice coverage of the globe continued for thousands of years and must have presented enormous problems of survival for all life forms. The ice did not begin to recede even from the equatorial regions until about 40,000 years ago and from midlatitudes some 10,000 years ago. Vestiges of that last great glaciation are still with us in the ice caps of the North and South Poles, though rapidly disappearing. It takes little imagination to visualize what life must have been like for these people, living continually in a climate like that in Antarctica today but without all our modern comforts and amenities. Social gatherings around a fire to hear stories of the ancestors and make music must have been a significant social bonding experience for these people just as it is for indigenous peoples today.

Homo sapiens emerged just before this last glacial epoch but managed to survive through it. In archaeological terms, this is described as the Palaeolithic or Old Stone Age Period because earliest Man used stone tools and was a hunter, killing animals as his principal source of food. No precise date can be given for the end of this period, as Man developed in different ways in different regions of the globe. But a convenient point at which history, as opposed to pre history, is taken to begin is at 10,000 years ago (ca. 8000 BCE). This time is sometimes designated as the start of the Neolithic or New Stone Age. It is marked by, amongst other things, the use of polished as well as chipped stone tools, and by the appearance of pottery that is often decorated with socially or spiritually significant symbols. This was the period of the beginning of agriculture and the domestication of animals in the Middle and Far East and in Central America. Some indigenous peoples, like the pygmies in central Africa and the aborigines of Australia, still live by an essentially hunting existence today. From such people we get at least a glimpse of what life might have been like for early Man. Because the sun, the moon, the stars, the elements of the weather were all associated with the sky, the first gods are believed to have been sky gods, particularly those of sun and moon. Such a god has been found by anthropologists amongst the pygmies and aborigines. As hunting was often done at night where possible, so that the quarry might be asleep and a more vulnerable target, it is thought that the moon god, or more often goddess, came first. As tribes adopted agriculture in the Neolithic, the food-producing labour was done in the day-time, so the masculine sun god gradually replaced the moon goddess. Some pagans today are bitheistic paying homage to both a god and goddess, while others are polytheistic, paying homage to various natural elements. The Japanese religion of Shinto is similarly polytheistic, paying homage to many nature gods or spirits. The gods were embodied with all those qualities of omnipotence and omniscience that Man aspired to but which remained forever beyond his reach. The gods were cultural ideals but were also believed to have temporal power.

When the hunter-gatherer became the farmer, after the agrarian revolution some 8000 to 10,000 years ago, the primary gods of sun and moon began to be replaced by the Earth Mother as protective goddess. In Mesopotamia she was known as Inanna while in Syria she was called Asherah or Anat. For Egyptians she took the name Isis, and the Greeks had several names for her: Demeter was originally the Earth Mother (*de* or *da* meaning 'earth' or 'good'; *meter* – mother; *da* still means 'good' in Welsh, and *meter* became *mater*, the Latin word for 'mother'), but Demeter later became the Greek goddess of corn. Demeter's daughter Persephone is also associated with growth in the natural world; Persephone was believed to spend the autumn and winter months, when plants were dormant, with her uncle Hades in the Underworld, to return again in the Spring. Both Demeter and Persephone are thought to date back to Neolithic times. Hera was a goddess also known to the Mycenaean culture in the Neolithic era and, as wife of Zeus and daughter of Cronos, she was to feature prominently as goddess of marriage and procreation in classical Greek culture. Aphrodite, goddess of love and procreation, was yet another image of the Great Earth Mother in ancient Greece. When Roman culture replaced that of the Greeks in the Mediterranean land, the Earth Goddess became known as Diana, guardian of the wilderness. When the Romans invaded Britain, they encountered the Celtic Earth Mother, Cerridwen. Indigenous peoples have always had a great respect for the feminine!

These goddesses were not demure creatures, endowed with the sort of qualities we think of today as associated with ladies of refinement! They were often vengeful and bloodthirsty, being equally associated with hunting and agriculture. Without the benefit of modern technology, farming in the Neolithic era was not a gentle occupation as farmers battled against harsh elements to till, sow and reap with primitive tools. Agricultural myths are pervaded by stories of the earth claiming life (animal or human) before it would bring forth crops. The Syrian goddess Anat was the wife and sister of Baal-Habad, the god of storms and fertility, symbolizing the struggle of agriculture, even in the fertile lands of the Middle East. There are stone tablets dating back to the 14th century BCE telling of the battle between Baal and Yam-Nahar, the god of rivers and the sea. In one legend, Baal slays a seven-headed dragon called Lotan (or Leviathan in Hebrew) who, in many cultures, represents formless chaos. Hence, Baal was important as a god in biblical times, so much so that the name tended to be associated with any saviour god.

As we move from the period of pre-history to what we call historical time, we can assign dates to events with greater precision, or at least with a far lower degree of uncertainty. One of the reasons for this is the appearance of early forms of writing, or what archaeologists call protoliterature, amongst Mesopotamians some 5000 – 4000 years BCE and somewhat later in Asian civilizations. The earliest Western writings were in cuneiform script — letters formed by pressing a wedge-shaped stylus into wet clay tablets. These records have been found amongst Babylonian, Assyrian, Hittite and Persian civilizations of Asia

Minor. This was the period in which the first cities were constructed.

The archaeological sequence of Man's development continues with the appearance, some 4000 – 3000 BCE in Asia and the Middle East, of the working of copper and bronze (an alloy of copper and tin) to usher in the Bronze Age. Although iron, possibly derived first from meteorites, was also used at this time in much the same way as copper and bronze, distinctive iron working did not appear until 1400 BCE, lasting until about 700 BCE, so this era is called the Iron Age. This working of natural materials into adornments for the body and spiritual effigies is symbolic of the union between man and the natural world.

Thus the pre historic and early historic cultural sequence of Man is through the Old Stone Age (Palaeolithic: 1 million to 10,000 years ago), New Stone Age (Neolithic: from 10,000 years ago or 8000 BCE until 4000 BCE), Bronze Age (ca. 4000 1400 BCE) and Iron Age (ca. 1400 700 BCE). This period embraces many great civilizations in the Mediterranean region, such as the Minoans, Mycenaeans, Hittites, Persians, Phoenicians, Assyrians, Egyptians and Sumerians, to be followed in Europe by the ancient Greek civilization. These dates are only approximate since the stages of civilization vary in their times of development between different regions, and even between Britain and continental Europe. Each group had its own ideas of social organization and agriculture, each developed its own form of writing and the groups do not appear to have influenced each other greatly at the start. The mutual interaction increased with population as human civilizations developed. What the history of these societies teaches us is that, while some civilizations were conquered by others, others simply died out because they were over-populated or over-exploited their natural resources.

The small community of the Easter Islands, known locally as Rapa Nui, present a salutary lesson for contemporary society. It is thought that the islands were first colonized around 300-400 CE by Polynesians. They lived successfully within this community for a thousand years with a population, it is believed, reaching 10,000 or even 15,000 by the 15th

century. But this was too large a population to be supported by the limited ecological resources. As a result of extensive deforestation and over-population, civil war decimated the community so that by the 18th century, when Europeans discovered the islands, only some 2000 inhabitants remained. As usual, western 'civilization' brought with it diseases unknown to the islanders so that soon only hundreds remained. As anthropologist Clive Ponting has said: 'The fate of the Easter Islands is a symbol for the overall population of the earth' [298].

The historical developments within the Sumerian civilization are of immense importance because, between 600 BCE and 600 CE, they led to the foundation of the three great religions of the West – Judaism, Christianity and Islam. The Books of the Jewish Bible or *Tenakh* (what Christians call the Old Testament) were transmitted orally for many generations and were put into written form sometime after the destruction of the Second Temple in 586 BCE. The name of Ezra (*fl.* mid 5th century BCE) is sometimes associated with the final written form of the *Torah* or Pentateuch. The *Tenakh* describes the history of the Jewish people and their traditions, with myths or legends woven around real historical people, places and events symbolizing their spiritual significance. It probably assumed essentially the form we know today by the time of Jesus. The degree of historical accuracy of the *Tenakh* has been a subject of controversy for biblical scholars for many years. As this book provides the foundation for western law and religion, it is perhaps the most significant collection of myths and traditions in the history of humankind.

The wisdom of the trees

As described above, many myths and traditions are associated with features of the natural world – trees, rocks, mountains and rivers – and the wisdom they were believed to hold. One of the definitions of 'wisdom' is 'spiritual awareness', so the use of this term to describe the behaviour and properties of trees in a spiritual world-view is not as inappropriate as it may at first seem [141]. Indigenous peoples who live in homes fabricated out of blocks of ice, or wood, or natural stone are at once closer to the natural world than those of us who live in homes fashioned out of synthetic bricks, breeze-blocks or plasterboard. The re-telling of myths by the elders of the group to the young, especially round the fire at night when the span of the universe is there for all to see in the star-strewn night sky, is part of this harmony with the natural world that we have largely lost: as Thomas Berry says, 'We have disengaged from that profound interaction with our environment that is inherent in our nature' [27, p.15]. Many of the stories in this section have been taken from the beautiful and inspiring book by Fred Hageneder [141].

Trees have been on Earth for more than 300 million years. Throughout human civilization they have been associated with magic and ritual because it was believed that they were imbued with a certain spirituality: and spirituality has always been associated with wisdom. As Karen Armstrong says: 'Trees, stones and heavenly bodies were never objects of worship in themselves but were revered because they were epiphanies of a hidden force that could be seen powerfully at work in all natural phenomena, giving people intimations of another, more potent reality' [7, p.17]. Because trees were usually much longer-living than humans it was believed that they retained knowledge from one generation to the next and, as a result, that they were home to the spirits of past generations. Even in our highly rational day-to-day existence, most of us feel a sense of enhanced spirituality when walking through woodland.

As well as the spiritual quality of trees, there are scientific experiments that show that trees and other plants can communicate with one another chemically through release of hormones when insect infestation threatens[1]. This is a response provoked as part of the genetic programming of the trees: it is not an example of intelligent reasoning or spiritual insight by the trees in the human sense, but it does illustrate the holistic nature of the plant-insect-environment ecosystem. But there is also some empirical evidence that plants may communicate by using energy fields [11; 369]. Pearl Weinberger and Mary Measures working in the Department of Biology in Ottawa University in the 1960s found a positive response from wheat subjected to certain sound vibrations; and there are numerous anecdotal tales of plants thriving as a result of tender loving care and kind words from gardeners, or by the playing of gentle music, while plants subjected to cacophony cringe from the noise. Experiments have shown some plants curling up in defence when threatened with attack by someone armed with a pruning shears [369, p.37], so trees may not be without a certain degree of spiritual awareness even in the human sense.

Trees, shrubs and plants of all kinds have always been admired as a source of spiritual uplift. The Egyptian pharaohs all had extensive gardens attached to their palaces and the wealthiest in Egyptian society had pleasure gardens on a smaller scale. There were few inhabitants in Roman society who did not have a garden attached to their home, both for the growing of herbs and for aesthetic pleasure[2]. That the use of gardens and window-boxes was widespread is confirmed by written sources, e.g. those of the 1st century Roman poet Martial in his *Epigrams* [239, section 11.18]. The ancient Greeks had their green spaces too, but mostly outside of the metropolis because running water was not generally available to Greek urban areas until after the Roman conquest. Beautiful gardens and courtyards embellished with fountains and running water were a feature of the homes of the caliphs and even public areas in Cordoba and Toledo at their zenith during the Moorish reign in al-Andalus. In the 21st century, gardening remains one of the most popular of hobbies and city dwellers often prize their window-boxes. The setting up of National Parks and Country Parks is a symbol of our increasing awareness of our need for green spaces that preserve our natural landscape, and psychologists believe that many of the behaviour problems of children today would decrease if they spent more time playing in green spaces.

Trees and shrubs and grasses have thus played a vital role in the social lives of all early cultures as they do for people today. The Celts are one group of early civilizations who venerated trees as repositories of knowledge and memory and the domicile of spirits. One possible derivation of the word 'Druid' is that it means 'wise oak', according to the Roman writer Pliny the Elder (*ca.* 24-79 CE) in his *Naturalis Historia.* Only the Druids were allowed to harvest the mistletoe that grew symbiotically on the oak, willow, rowan, maple or hazel, and the mistletoe, like the oak, had special spiritual significance. This must have seemed a particularly magical plant to early pagans as it grew high up on a tree but had no roots in earth. To kiss a maiden wearing a crown of mistletoe in her hair would bring good fortune, and for a couple to kiss under a bunch of mistletoe would bless their union – a custom we have diluted to preserve at Christmas time.

The old Scandinavian word '*vid*' means wood or forest but it has given us a number of words associated with knowledge or wisdom: *witan* (Old English: to know), *wissen* (German: to know), 'wits', 'wise' and 'wisdom'. We shall meet other examples of the association of trees and knowledge again shortly. The cypresses of Crete were revered by the Minoans, and they survived for centuries until over-exploitation by the Venetians during the Renaissance to provide timber for shipbuilding forced them into extinction on the island. Most of the so-called 'primitive' civilizations knew how to manage natural resources in their environment and live holistically with them. The eucalyptus tree that grows prolifically throughout Australia is, not surprisingly, associated with all kinds of sacred rituals for the aborigines. These are not to be shared with those outside of the tribes but are handed down orally from generation to generation by the tribal elders. The *yidaki* or didgeridoo is a long hollow-tube musical instrument fashioned from a eucalyptus branch that has been hollowed out by termites and it is used to accompany ritual dances. The low, deep note produced when the tube is blown into is said to produce resonating vibrations deep in the chest of the hearers, vibrations that are conducive to the establishment of a meditative trance.

The beech tree has a special claim to be associated with knowledge and wisdom. The beech together with the yew were the woods favoured for the creation of ogham sticks and runes, though the latter more commonly found are made of more durable metal or stone[3]. The ancient ogham and runic alphabets were line symbols carved on wood or stone that were used by the bards for passing secret messages to one another. They were also used for divination by Celts, and by Germanic tribes in central Europe and Scandinavia from at least 1200 BCE. In divination, the Druids would gather together a selection of ogham sticks in their hands, ask a question of the spirits, then cast the sticks or runes on the ground. The Druid would then make a prognostication depending on what combination of symbols were uppermost. The ogham sticks and runes are the European counterpart of the book of *I Ching* ('The Book of Changes') used from earliest times in the East and still in use today in China and Japan[4].

There were originally twenty ogham symbols or fedha, later increased to twenty-five, each representing a different tree and each associated with a season of the year. For example, the fir and silver birch represented the winter solstice, the oak the summer solstice; the rowan and willow symbolized Imbolc (end of January, early February, a celebration to stimulate the re-awakening of the earth); hawthorn for Beltain (end of April, early May, to celebrate the burgeoning summer and including the May-day festival); gorse and hazel for Lammas (end of July, beginning of August, to celebrate the harvest; in Britain, the August Bank Holiday used to fall in Lammas until it was moved to the last Monday in the month in 1971); elder and yew for Samhain (end of October and beginning of November, to mark the beginning of winter and including Halloween); and so on.

In Scandinavia and Germany, the different runes, of which up to three were selected in each throw, were associated with one of 24 Norse deities. Again, each rune combination or *raedan* (which gives us our English words 'reading' and 'riddle') had to be interpreted by the tribal seer. The etymology of trees, wood, knowledge and wisdom is frequently linked from our pagan heritage. In German, the beech tree is *die Buche* and the word for book, of the same origin, is *der Buch*, while a letter of the alphabet is *der Buchstabe*, literally 'beech sticks', indicating their ogham background. In modern Swedish the word *bok* can mean either beech tree or book. Of a different but still interrelated etymology we have the pages of a book often referred to as the leaves. The German word *das Laub* means foliage or leaves, related to the Welsh *llyfr* and French *livre* for book, and the Welsh *llythyr* for letter.

The birch tree was always associated with protective powers. It is reported [242] that London used to have a birch tree growing in the Strand that was used annually as a maypole until in 1717 Sir Isaac Newton had it cut down to use as a support for his newly invented telescope. Even the appliance of science needed a little magic to help it make significant discoveries; and Newton, one of the greatest of all English scientists, had a great interest in alchemy and magic.

The early pagans lived their whole lives in holistic ecological harmony with the changing faces of the Earth [77]. Samhain marks the beginning of the dark months of winter in the Northern hemisphere and is therefore associated with ghosts and goblins and the spirits of the departed. It includes Halloween on October 31st. This pagan festival was hijacked by the Christian church, as they did so frequently, and turned into All Saints Day on 1st November. Coincidentally, the armistice for World War I happened to be signed in early November so our Western commemoration of those who have fallen in battle still occurs in Samhain.

Specific trees that are native to certain geographical areas often become associated with local myth, folklore and spirituality. The baobab tree is venerated in Africa as the Tree of Life because its massive trunk can hold many litres of water – enough to sustain a small village for some days – so providing a very practical spiritual image. The indigenous peoples of New Zealand, the Maoris, believe that people and trees are both descended from their god *Tane Mahuta* ('Lord of the Forest'). If a tree has to be cut to be used for some construction work, a sacred ceremony has to be performed by the elders of the tribe. Particularly important are the native kauri trees, of the genus *Agathis*, which grow to a great height and girth over many hundreds of years. The kauri trees in New Zealand are the giants of the forest, like the redwoods of California.

North American Indians refer to trees as 'our standing brothers and sisters'. The monkey-puzzle tree, *Araucaria*, is so venerated by a native tribe in Chile that they take their name, Pehuenche, from the tree (*pehuen* – monkey-puzzle; *che* – people). The tree is used as a food source for the tribe with bread made from ground seeds, while the resin is used medicinally. Charles Darwin encountered a solitary sacred acacia tree in Patagonia in 1833 that the local natives called the Altar of Walleechu: the locals hung offerings on the tree to keep away the evil spirits that Walleechu represented. The tree was situated at a geographical location of some spiritual significance near the confluence of the Rio Negro and Rio Colorado rivers. The Garden of Eden (or Edeen as it is called in the Epic of Gilgamesh) was located at the meeting place of the two great rivers, the Tigris and Euphrates to give the land known as Mesopotamia (Greek: *meso* – between; *potamos* – river).

The 'rod of Aaron' appears in the Old Testament endowed with many miraculous powers [Exodus 4: 2-5; 17: 5, 6]. This was a staff cut from the almond tree [Numbers 17: 7, 8]. The staffs of the three wise men and the crozier of the high clergy today are supposed to be symbolic remembrances of the rod of Aaron. Amongst the Sumerians, the cedar tree was held to be the abode of their god Ea or Enki, Lord of the Earth, and god of wisdom, so the tree became a location for prophecies and oracles. The cedars of Lebanon appear in the Old Testament as symbols of power and longevity [see, for example, Ezekiel 31: 3-9]. The roof and walls of the Temple of Jerusalem were made from panels of wood from the cedars of Lebanon. Olive wood was used for the doors and two cherubim on either side of the altar were overlaid with gold [1 Kings 6]: the cedar symbolized power, the olive wood peace.

Moses was instructed by God to make a fragrant oil from myrrh and cinnamon and use it to purify the Ark of the Covenant and altar in the tabernacle and for the anointing of priests [Exodus, 30: 22-31]. The three wise men are said to have brought gifts of frankincense and myrrh to the baby Jesus, together with gold, considered to be the purest of metals and always a symbol of the sun god [Matthew, 2: 11]. Frankincense and myrrh are sweet-smelling resins of tropical trees, and were highly prized and often traded in earlier times, especially in the Middle East.

In Islam, the olive tree is venerated and mentioned in the Koran in the *sura* devoted to Allah as the Light of the world: God is the light of the heavens and the earth. His light may be compared to a niche that enshrines a lamp, the lamb within a crystal of star-like brilliance. It is lit from a blessed olive tree neither eastern nor western. Its very oil would almost shine forth, though no fire touched it. Light upon light; God guides to His light whom He will [*sura* 24: 35].

Other trees have significance in other contemporary religions. The banyan tree, one of the *Ficus* (fig) genera, is linked with *Brahma*, creator of the universe in the sacred Hindu scriptures, the *Vedas* and *Upanishads*. The banyan is therefore the Tree of Knowledge. One of the reliefs in the Stupa of Barhut, built between 184 and 172 BCE, shows the enlightenment of the bodhisattva Kasyapa as occurring while sitting under a banyan tree. The enlightenment of the Buddha himself is said to have occurred while he sat under a pipal tree, another of the fig family, *Ficus religiosa*. Before effigies of the Buddha started to appear in the 2nd century CE, the Buddha was represented by a wheel (indicating the unity of all that is) or a pipal tree. The pipal and banyan are revered throughout Asia, as is the ginkgo tree. After the atomic bomb was dropped on Hiroshima in 1945, every living thing within several miles of the epicentre was destroyed – except for four ginkgo trees, the closest less than a mile away from the epicentre, that survived and began to blossom again in the following spring. All four trees still flourish today and, not surprisingly, the ginkgo is therefore regarded by the Japanese people as a symbol of hope. In western herbal medicine extracts of the ginkgo are used to treat dementia and circulatory problems.

In classical Greece, the ash tree, *Fraxinus*, was the symbol of the god of the sea Poseidon, who became known as Neptune in Roman times. Pieces of ash wood were taken on sea voyages as protection for the travelers. This custom remained even until the 19th century when European migrants crossed the Atlantic Ocean to America. A Druid's staff, dated back to the 1st century by radiocarbon dating, and made of ash-wood has been found on the island of Anglesey. It bears a solar spiral at the crook, from which it has been associated with the Welsh wizard Gwydion, nephew of the great magician Math. The Icelandic book of legends, the *Eddas*, describes how the first man was made of ash, and the first woman of elm – no more fanciful surely than having the first woman made of a man's rib!

The holly tree (*Ilex*), especially made into a wreath, has become associated with Christmas because its spiky leaves are symbolic of the crown of thorns that Jesus was made to wear for the crucifixion. It also bears its bright red berries at that time of year symbolizing new growth and, with it, new hope. The old High German word for holly is *hulis*, meaning 'holy'; modern German uses a much more onomatopoeic word – *Stechpalme*. The elm tree was considered in English folk-lore to be symbolic of the meeting place of mortals and spirits. It is interesting to note that so-called 'Dutch Elm Disease' destroyed elm trees throughout Britain at a time when spirituality was also declining, considered by some to be symbolic. The German word for the elm is *Elfenholz* which, when translated, gives us the English name 'elfin wood' for the elm.

The linden tree, *Tilia*, was the focus of village life in Germany even into the 18th and 19th centuries. A tea made from the flowers of the linden tree has long been used as a tonic for the blood, reducing cholesterol levels and lowering blood pressure, centuries before these physiological measures assumed the importance they have today in fighting arteriosclerosis. The soothing qualities of linden extracts, and the prominence of the tree in German social life, gave rise to the German word *lindern*, meaning to soothe, to alleviate.

Since the time of the ancient Greeks at least the longevity, the girth and dark density of the yew has been associated with death and transfiguration into immortal soul. Of the several gargantuan yews in English churchyards, there is one at Much Marcle in Herefordshire that is nearly 10 meters in circumference and, like the yew in Tennyson's poem, is thought to be over 1000 years old.

The seasons bring the flower again,
And bring the firstling of the flock,
And in the dusk of thee, the clock
Beats out the little lives of men.

O, not for thee the glow, the bloom,
Who changest not in any gale
Nor branding summer suns avail
To touch thy thousand years of gloom
Alfred Tennyson, *In Memoriam*, Stanza II

The association of the yew tree, *Taxus*, with Samhain was mentioned above. Samhain was the time when the gates between the worlds of the living and the earthly departed were open. This is why so many graveyards to the present day have yew trees growing within them. It also probably has something to do with the fact that most parts of the tree are very poisonous. Making use of this property pharmacologists found that paclitaxel, from the bark of the yew tree, was a powerful anti-cancer drug. Because of their longevity through successive human generations, yew trees are also associated with the continuity of the life process after earthly existence has ceased.

In September of 1842, Charles Darwin moved into a new home, Downe House, in the village of Downe, near Bromley in Kent. He would often sit to think in the shade of the great yew tree in St. Mary's churchyard for inspiration and it was his wish to be buried there, in peace and tranquillity, but the nation decreed he must have a more fitting memorial in Westminster Abbey.

The aspen is venerated by several tribes of North American Indians, particularly the Navajo and the Cheyenne. The Hopi Indians smoke leaves of aspen on ceremonial occasions. In ancient Greek mythology, Heracles wove himself a wreath of aspen leaves and lit a fire made from aspen wood as an offering to Zeus for his safe return from the underworld. The aspen is associated with the warding off of evil spirits. The old Irish *Sanas Chormaic* ('Chronicles of Cormac') says: 'a rod of aspen was used by the Gaels for the measuring of the bodies and the graves in which they were interred; and this wand was always in the cemeteries of the heathen'.

The juniper tree was regarded by Germanic pagan tribes as a watchful sentinel, no doubt because of its erect habit. It was an intermediary between the mortal and spirit world. The modern German word for a juniper, *der Wacholder*, reflects this (German: *wachen* – to be awake, alert).

In Egypt, archeologists have found wooden boats used as burial vessels in place of more conventional coffins. There were at least a dozen boats buried at the royal burial site at Abydos, located about 450 km south of Cairo. The wooden planks of which the boats were constructed were lashed together with rope. The boats were probably for a Pharaoh to use in the afterlife as he sailed down the Nile with the sun god, Ra. One boat is some 25 metres long and is thought to be about 5000 years old. There were also five boat burials discovered near the Great Pyramid of Cheops (Khufu) at Giza of about the same age. To the Egyptians the boat journey symbolized the soul's journey to the afterlife.

Though less well known, there is a Woodhenge located some 30 kilometres from Stonehenge at Durrington. Although its existence was discovered in the early part of the 20th century by aerial photography, it was not until 1999 that it was excavated by archeologists. The red cedar circular construction, aligned with the mid-summer sun, dates from before 2000 BCE. It was probably used at some periods of the year as a sacrificial temple, for the body of a three-year-old child with a deep groove in its skull was found within the circle.

This short selection of myths and traditions of spiritual and religious practices amongst the people of earlier ages will give some indication of the extent to which trees, keepers of secret knowledge since they first appeared on Earth, have interacted with humankind. Their wisdom lies in their potential to provide so many of the needs of humankind, both physical and spiritual, and in the fact that they have evolved in a balanced ecological system that we are now flouting.

The tree may be viewed as an allegory of the human individual – the trunk represents the individuated self, the roots are the ancestors and their traditions from which the self develops, the branches are the connections we make with the physical and spiritual world, and the leaves that are shed each autumn to nourish the ground beneath are the thoughts and ideas that we disperse to nourish humankind as a whole.

The seer Margaret Silf has presented a lovely metaphor for human behaviour, as it should be, in the message of wisdom she intuited from her encounter with an oak tree:

> how it is better sometimes to bend than to break;
> how it is necessary to let the blossoms fall, to make way for the fruit, and to let the fruit fall to nourish the creatures of the forest, and to strengthen her own roots;
> how the leaves she grieves to lose become the blanket of her winter sleep, sheltering her dream until the sun returns;
> how she quivers with joy when the children play among her boughs, and how she shudders with dread when the lightning strikes, and how both are part of life and ask for our consent;
> how the fresh leaves at the tips of her branches have no idea of the vast structure of life, of boughs and trunk and roots that hold them in being and bring them to their awakening, but how they awaken

anyway, without needing to know why and how;
and how good it is to be, and to grow, and to trust.
Margaret Silf, *The Way of Wisdom*, Lion-Hudson 2006, p.35

Wood eventually decays but stones are much more durable. This is why stone monuments have been constructed for thousands of years by those who wanted to leave an almost indestructible record of their existence and why stone constructions were popular with early man as a place to pay homage to the gods.

The message of the stones

I made these verses in a rocky land,
And I have named them Voices of the Stones,
Although they do not keep that innocence
Was shed on me when quiet made me kin
To the cold immobile herd. All things have changed
From primal nature save these stones: all things
Since Eden, bird and beast and fin, have strayed
Far from that shining garden of His thought:
We also. Only the humble stones have kept
Their morning starriness of purity
Immutable. Being unfallen they breathe
Only unfallen life: and with my cheek
Pressed to their roughness I had part regained
My morning starriness, and made these songs
Half from the hidden world and half from this.
Voices of the Stones, by A.E. Russell [301]
© The Macmillan Company, New York, 1925

Indigenous peoples have always lived at one with their natural environment and with the spiritual domain, perpetuating traditions handed down through many generations in myths and legends. We have seen something of the veneration for the wisdom of the trees. But all of Nature – woods, rivers, mountains and standing stones – were believed to hold the essence of the cosmic spirit.

The stone monuments the early peoples erected suggest that a kind of nature religion, involving worship of the sun and moon, was always an important feature of their lives. Standing stones or natural rock formations have particular spiritual significance because they are unchanging during the span of a human lifetime – or even several lifetimes – unlike other natural features that change with the seasons. The relative constancy of rivers also gives these features a special place in spiritual ritual. Earthly objects were, in some sense, symbols of the greater spiritual force that controlled the seasons, night and day, the wind and the rain. The word symbol is derived from the Greek, *symbolon*, and *syn* + *ballein* meaning 'to throw together' – something that brings together the material and the mystical, uniting the physical with the spiritual in holistic union and serving as a metaphor for a social 'bringing together'.

As part of this spiritual empathy, communion with the ancestors of the tribe has special significance. Sacred sites were established where this communion took place, like Ayers Rock in Western Australia, known to the aborigines as *Uluru,* which has now become a well-known tourist attraction. These sites were believed to be invested with a special spiritual energy. The Australian aborigines call the time of the ancestors the Dreamtime or *Alcheringa*, which they visit in trance-like prayer states or in dreams, often to seek guidance in their daily lives. It was Freud who first drew attention to the significance of dreams in reflecting our subconscious memories and desires [107]. If we are not undergoing psychoanalysis, we are unlikely to pay much heed to dreams in western civilization, though they are considered of great significance in Sufism and in many indigenous cultures. To the Australian aborigines and the North American Inuit, dreams exert a great influence on their waking lives.

Aborigines have a passion for painting traditional spiritual images on rocks and on tree bark. The imposition of western values that are destroying their native culture the aborigines call The Great Forgetting. The oldest rock paintings in Arnhem Land, in the Northern Territory of Australia, have been dated by anthropologists at some 40,000 years old. The kinds of images aborigines paint today are highly reminiscent of those they painted at the dawn of their civilization. These 40,000-year-old paintings show the didgeridoo in association with traditional tales of these people, represented on the cave walls. Music was used to accompany rituals in prehistory just as it does for indigenous peoples today. Music and dancing are associated with courtship rituals for young people in present western culture, and music is similarly used to accompany dramas on television or in the cinema to enhance immersion of viewers in the story.

With a Divine associated with the heavens, mountains became particularly spiritual places as worshippers could ascend that much nearer to heaven. The 'God in Heaven' image has a very long history. Thus in holy scriptures we are told that God gave the Ten Commandments to Moses on Mt. Sinai [Exodus, Chap. 20]; Jesus went up into Mt. Hermon for his transfiguration [Matthew 17, 1-8]; Muhammad went up into the hills above Mecca to receive the words of archangel Gabriel that became the Koran [*Qu'ran*, *surah* 2, *Al-Baqarah*]; and Joseph Smith went up into the high ground of Mt. Cumorah to receive from Moroni the Book of Mormon. Such high ground is a symbol of humankind's belief that there is a 'higher' spiritual domain that exists above and beyond earthly material existence. When Sir Edmund Hilary and Sherpa Tensing reached the summit of Everest in 1953, Hilary planted a Union Jack to claim it for Britain while Tensing fell on his knees and offered up a prayer to the mountain asking forgiveness for invading its sacred terrain.

Megaliths of natural stone have been built from the earliest civilizations to the present day as monuments to the lives of great rulers, both temporal and spiritual. The pyramids of Egypt that were built as tombs for their god-like pharaohs and the flat-topped temples of the native tribes of the Americas are well-known examples of such constructions out of native stone. We still have the Great Stupa of Borobudur in Indonesia and the Parthenon of Greece as temples to the gods. Amongst Aztec and Inca ruins there are many stone constructions, the purpose of which is unknown, but the likelihood is that they were associated with worship or ritual, and possibly sacrifice in some way. The Incas also had their *Huacas* – mountains, streams and natural rock formations that they believed were imbued with spirituality. The great Mayan pyramid of Kukulcan at Chichen Itza has 91 steps on each of its four sides, representing days of the four seasons, and one final four-sided step at the top, giving 365 steps in all, clearly of some significance in measuring out the year. The Inuit of North America have lived in their *Nunatsiaq* (beautiful land) for over four thousand years constructing *inuksuit*, structures of natural stone as loci of spirituality that celebrate a deep and total attachment to the land. In the present day, the red rocks of Sedona, Arizona, are a place of pilgrimage for mystics who believe that the region has a concentration of energy vortices.

There are many examples throughout Europe of megaliths constructed by early peoples to celebrate natural events, especially those involving cycles of the sun or moon, to which were attributed religious significance. Lyall Watson describes [378, p.30] how the standing stones known as the Rollright Stones in Oxfordshire in Britain emit a low frequency sound at certain seasons or times of day that can be heard when standing near, but nothing can be heard within the circle of stones [317]. These sounds have been investigated by scientists, but no convincing explanation for their occurrence has been put forward. Solid state chemist Don Robins suggests tentatively that it may have something to do with irregularities in the crystal structures of the stones [318].

On Salisbury Plain in Wiltshire in Southern England there are stone circles that have been the focus of attention for spiritual people (and, more recently, for tourists) for many centuries: these are the constructions of megaliths at Stonehenge and Avebury. These are thought to date from before 3000 – 2000 BCE. They were probably used by the Celtic Druids later, just as they are used by Neopagans now, but their origin almost certainly lies well before the Celtic immigration into Britain[5]. The erection of such massive constructions as these must have involved considerable time and labour, which must have been thought worthwhile for their spiritual ends – an indication of the importance of spirituality in the lives of early peoples.

In Scotland, the ceremonial sites are much less accessible than those of Wiltshire. The two most important are the Machrie Moor Standing Stones on the Isle of Arran and the Callanish Standing Stones in the Outer Hebrides. At Machrie only four stones now remain of two large circles. Erected at a time when the climate would have been much more hospitable and welcoming to the Celtic immigrants, the Machrie stones guard the graves of Neolithic farmers whose skeletal remains have been found buried with their arrowheads. The Callanish stones in the Isle of Lewis still form a veritable petrified forest which, legend has it, represent thirteen giants who refused to convert to Christianity and were turned to stone as a consequence by St Kieran in a particularly uncharitable act.

In Wales, the Pentre Ifan Cromlech near Nevern in Pembrokeshire, known locally as Arthur's Quoit, served as a burial ground for up to fifty tribesmen and women. It consists of a curved cap-stone, nearly 5 meters long, supported by three uprights. Some time after its construction the burial ground was covered with a mound of earth to make a barrow, rather like that at Silbury Hill, bounded by a dry-stone and timber wall that has disintegrated. Other Celtic burial sites at Samson's Quoit and Llech y Drybedd ('Stone of the Three Graves') are nearby. In the Presceli Mountains of Pembrokeshire there is a hillside, Carn Menyn, that is strewn with rough-shaped weathered boulders that have fallen from the rock formations above, like some gigantic scree. Perhaps the original site of these megaliths was itself of spiritual significance to make the onorous task of carting the stones to Wiltshire for Stonehenge worthwhile, or even essential, if the new site was to be spiritually blessed.

Some of the best known examples of stone tombs or dolmens amongst these early Celtic pagan settlements are to be found in the Eastern part of Ireland, with the Newgrange megalithic tomb the outstanding example. These structures probably date back to before the pyramids of Egypt, more than 3000 years BCE and a long time before the expansion of the European Celts into Britain. They were probably used both for burial of the elders of the tribes and as temples to the sun and moon. These testimonies to pagan religion have survived in the more remote corners of Europe where the Roman invaders were never able to fully subdue the local population and form settlements of their own.

It is not only in Britain that such ancient stone monuments survive. In Sine-Ngayene in Southern Senegal the necropolis is marked with a series of several dozen circles of small standing stones, each less than a metre high. The site here is also of Iron Age origin. At Tagarp in Sweden, an enclosed tumulus comprising passage and burial chamber within has a memorial site outside for worship so that successive generations could come and pay their respects to the ancestors. Such megalithic tombs are to be found in many other sites across Europe, but the menhirs and dolmens at Carnac are by far the most impressive. Carnac is a small town on the southern Atlantic coast of Brittany in France where there are more than 3000 Neolithic monuments of various kinds. The intuitive Natasha Hoffman has tapped into the divine spiritual energy through her spirit guides she calls archangels, using the sacred stones at Carnac as a conduit [174].

According to Hoffman it is the pollution of the human mind with materialism that has led to the worldwide physical pollution of the planet. 'False materialistic goals posing as progress, and a lack of awareness of the vital significance of the cosmic energy patterns' have impeded our harmonious spiritual development [174, p.34][6]. We are given a timely reprimand: 'For thousands of years you have lived without honoring natural laws. You have replaced spiritual law with man-made laws that are based on power and control; they are unrelated to the natural harmonies of the universe' [*ibid.*, p.45] and 'There will be a new culture based upon a relationship with other realms and celebration of the rites of passage of the soul's journey. . . These observances will free the spirit, as opposed to past religious doctrine that has imprisoned it in teachings of judgment and punishment. There will be a reawakening to the truth' [*ibid.*, p.85].

Often these lines of sacred stones follow lines of Earth energy, not so much electromagnetic as spiritual it seems, since they are more easily detected by dowsers than by the most sensitive of scientific electromagnetic detection equipment. Such lines were called ley lines by Alfred Watkins (1855-1935), the man who 're-discovered' them in 1921.

Trees and rocks are sacred and evolve just like the animal kingdom. Using natural stones as a conduit, especially those erected at sites where spiritual energy is concentrated along ley lines, is not unusual for mystics. The crystal ball used by a medium is simply a focus for the concentration and perhaps an aid in channelling the energy from the universal and divine spiritual field – which is why this is called 'divination'. Spiritual healers similarly often use the regular atomic structure of mineral crystals to channel healing energies.

> Beautiful and strong is the material of stone, but more beautiful and
> Much more powerful is the mystery that emanates from them.
> Li T'ai Po (701-762)

In a similar way, the standing stones erected on ley lines serve as a conduit for the universal spiritual energy to and from the Earth, just as the *chakras* and meridians serve a similar purpose in the human body to renew our *chi*. Ancient civilizations knew this and built sacred monuments of stone, partly to channel the cosmic energy and partly because such structures were more durable to serve as temples dedicated to the cosmic spirit than if they were made of wood. The stones themselves represent a part of the cyclic evolutionary processes of the physical Earth, just as living organisms are a part of their own evolutionary succession. All rocks and crystals have been formed, either out of the skeletons of living creatures of the Earth or out of the material of the Earth itself as it passes through its cyclic changes. Geological formations are a reminder of the continuing evolution of Earth.

The message of the stones is that we should not forget our ancient past and the important part that spirituality played in everyday life for our ancestors, as it does for indigenous people today: such people show great reverence for the workings of the natural world and live in harmony with them. The lives of less sophisticated peoples were regulated by the seasons, by the rising of the sun and the phases of the moon. There is an ancient wisdom encapsulated in an invocation to the Great Spirit used by the North American Indians:

> 'let me learn the lessons you have hidden in every leaf and rock'.

Summary

Tales of the gods or nature spirits form part of the mystical or religious dimension of holistic existence and, with it, contribute to the cohesion of a society. Some of the tales invoke national heroes; others portray persons embodying timeless moral messages. Myths would not have survived for the many thousands of years of human existence if they did not serve some practical purpose. Primitive people in particular would have had so many urgent practical matters to attend to that they knew affected their very survival, like the search for food and shelter and dealing with wild animals or rival tribal groups competing for the same resources. They believed that communing with the spirit world of past and present through the telling of stories and the practice of rituals was important for their survival. For us today these same myths and rituals represent the traditions of our respective cultures. They are socially cohesive and give us a sense of our history and our identity and a connection with the spiritual domain that constitutes an indispensable part of human existence. For children, fairy tales represent these same timeless moral messages in a language they can understand and at the same time stimulate their imagination.

While myths and traditions may seem largely irrelevant to us in our materialist existence today, a comparison of the artefacts reflecting human development unearthed by archaeologists has shown that there were similar practices in different parts of the world in similar time periods, with cultures communing through the cosmic spiritual field. There is also similarity in many of the archetypical symbols that are displayed in their artwork or that have been found in these archaeological artefacts, indicating an exciting and timeless spiritual connection between the myths and traditions in all cultures.

In this chapter, I have tried to capture something of the cosmic spirituality experienced by people through their cultural myths and traditions, and expressed through the trees and rocks of the natural world, and how their message has been an inspiration to countless generations – a message we need to heed so urgently at the present time. I have also indicated in Chapter 7 what rational scientists believe will be the consequences to humankind if we do not adopt this holistic world-view.

CHAPTER 5

SOCIETY

The quest for holism, by its very nature, must involve a cohesive society. By contributing what we can as individuals in living a holistic life, we set an example that we hope others in society will want to follow. If we look rationally at the way we live now it is obvious that we cannot go on pursuing our own goals indifferent to the needs of others, accumulating wealth to outshine our peers, demanding our rights while neglecting our duties, being profligate with Earth's resources and disrespectful of its organic coherence. This chapter explores some sociological issues that we need to bear in mind in our quest for the holistic life.

Living holistically requires that we are free. An individual's sense of being part of a society while having freedom to act as an individual has not emerged without the vision of a few thinkers who had the determination to fight against oppression by Church and State. It is the overwhelming power of these institutions and that of other organizations, like commercial corporations, the National Rifle Association, or the evangelical church, to cite just three examples relevant to 21st century America, that pose the greatest threat to the freedom, security and wellbeing of all [267]. Social problems in America have a way of migrating to Britain and elsewhere in a very short time.

History and sociology are inextricably intertwined: the past influences the present, and the present shapes the future for us all. Just as the study of history needs an appreciation of the social conditions prevailing at the time of great changes of the past, so an understanding of the human quest for freedom as individuals and societies needs some knowledge of the historical background leading to change. History is really an account of the changes in society brought about by powerful men and women. It is relevant therefore to include some brief biographical details of the lives of those social reformers who were most influential in gaining the freedom of the people and who laid the groundwork for the formal study of social behaviour.

This quest for freedom from oppressive rule of Church or State is an ongoing human saga that continues to this day and too often it has exacted a price in human blood. Martin Luther's action at the castle church of Wittenberg in 1517 in nailing his 95 theses criticizing the Catholic Church for simony and the sale of Indulgences, amongst other transgressions, was inspirational in provoking the Reformation. This marked a huge change in world-view in the west as it began to break free of the stifling authoritarianism of the Catholic Church that had held sway for 1500 years. It was a revolution in religious thought comparable to the secular renaissance of learning that had begun in Italy some two centuries earlier. The world was now ready to accept the discoveries of Copernicus, Galileo and Newton. Science would not have developed, or at least its progress would have been delayed, if the power of the Catholic Church had not been undermined. Henry VIII saw a series of Acts passed by Parliament in 1534 to abolish papal authority and establish a national Church of England. All monasteries in Britain were dissolved in the three years beginning in 1536 and, with them, much of the learning of the ancients. Now it was time for humankind to begin the quest for freedom for the people from the authority of aristocracy, Church and the monarchy. It is with these actions that people began to see themselves as a society rather than just as individuals and act accordingly.

The quest for power

An individual's quest for power has always been the driving force in producing change in society, whether or not they are pursuing philosophical ideals. As pointed out by philosophers from Thucydides to Hobbes, there are also personal and individual motivations behind their actions. Thomas Hobbes (1588-1679), from Malmesbury in Wiltshire, was an English philosopher whose writing focused on issues of politics and sociology. He was also the translator of the works of the Greek historian and social philosopher Thucydides (*ca.*460-*ca.*400 BCE) with whose philosophy he empathized. Hobbes was a fine linguist who could read and write Latin, Greek, French and Italian as well as his native tongue. Summing up the essence of his and Thucydides' common view of humankind's will-to-power Hobbes wrote: 'I put for a general inclination of all mankind a perpetual and restless desire of power after power, that ceaseth only in death. And the cause of this is not always that a man hopes for a more intensive delight, than he has already attained to; or that he cannot be content with moderate power; but because he cannot assure the power and means to live well, which he hath present, without the acquisition of more' [172, p.123]. There is much of this philosophy in the acquisition of material wealth today, as reflected in Alfred Adler's psychological evaluation of human motivation (Chapter 2). Hobbes saw both Church and State as the embodiment of ultimate power, portrayed by the metaphor of Job's sea monster, leviathan, that devoured everything in its path. Today, we need to add the mega-corporations to the list of those who hold supreme power, often more extensive than most Church or State hierarchies have ever achieved, even if less overt. Hobbes' *Leviathan* was published in 1651 during the British inter-regnum of Oliver Cromwell, which is just as well given its anti-royalist tone. The English Civil War of the mid-17th century marked Britain's attempt to assert the voice of the people through parliament against intemperate monarchy. Hobbes was also branded an atheist because of his opposition to the authority of the Church.

Life, liberty and the pursuit of happiness

An even greater social reformer of the Enlightenment period, a successor to Hobbes and one who expressed the same opposition to unbridled power, was the Englishman, Thomas Paine (1737-1809). Paine championed both the individual and society of the common man as a whole against both Church and State in their attempt to make all those in their power conform. Paine was born in Thetford in the East Anglian county of Norfolk. His father was a Quaker in belief and a maker of women's corsets by trade, an occupation that young Thomas at first followed. But Thomas was restless and ran away to sea in 1756. He served for only a year however and then returned to his trade of stay-making, first in London, then Dover, then Sandwich where, in 1759, he married a Mary Lambert who died the following year. Mary's father worked for the customs and excise, so Tom followed this profession over the next few years at Grantham, Lincolnshire, and Lewes in Sussex. It was here that his interest in politics began and here also he married for the second time. He wrote a pamphlet to try to improve the lot of excise men, but this and other factors resulted in his discharge from the excise service. A fellow excise man in London introduced Paine to Benjamin Franklin who persuaded Thomas to emigrate to America and even went so far as to write him a letter of recommendation.

Once he had arrived in America, at the end of 1774, a friend offered Paine the chance to become Editor of a publication called the *Pennsylvania Magazine*, which gave him the perfect opportunity to voice his rather radical views. In his advocacy of the setting up of a fund to help youngsters at the start of their careers and for the elderly who were no longer able to provide for themselves, Thomas Paine's efforts could be regarded as the inception of the ideals of the welfare state.

At this time, relations between England and America were already strained, exacerbated by the Boston Tea Party event of December 1773. The colony was rebelling against rule by an ineffectual English king, George III. In April 1775, Paul Revere warned his fellow citizens of Boston that English reprisals were imminent and in January 1776 Paine published his pamphlet *Common Sense*, which stated: 'To the evils of monarchy we have added that of hereditary succession' and talked of 'the distinction of men into kings and subjects . . . for which no truly natural or religious reason can be assigned.' Paine advocated 'securing freedom and property to all men, and above all things the free exercise of religion according to the dictates of conscience.' This and much more roused the already incensed American people and the Declaration of Independence from English rule was signed on the 4th July 1776 by the thirteen United States of America[1]. In declaring independence the Congress decided that the people had 'unalienable rights' to life, liberty and the pursuit of happiness. It was Thomas Paine's pamphlet, *Common Sense*, that was the precipitant factor, though the writings of English empiricist philosopher John Locke (1632-1704) did much to influence the content of the American Constitution that was drawn up between 1785 and 1791 (the Bill of Rights, with Amendments to the Constitution).

John Locke's advocacy of individual freedom and liberty was predicated on the assumption that humankind is predominantly rational in the motivation of its actions: 'The freedom . . . of man, and liberty according to his own will, is grounded on his having reason, which is able to instruct him in that law he is to govern himself by, and make him know how far he is left to the freedom of his own will' [224, Book II, para.63]. That was probably an overly optimistic premise. Humankind and, regrettably, more so men than women are driven first by the passions – not the gentle, caring emotions that consider the needs of others but, too often, the will-to-power that Thucydides, Hobbes, Adler and Nietzsche have all seen as the primary motivator.

By the early 1780s Paine was in France, which he had visited more than once during the previous decade trying to raise funds to help America. Having sparked the American Revolution, Paine was also a key protagonist in the Fall of the Bastille on 14 July 1789, which represented the key event of the French Revolution. In America, meanwhile, Paine was active in efforts to abolish slavery. These were not directly successful but his humanitarian stance was almost certainly one factor in provoking the American Civil War (1861-1865) between the North, who wanted to abolish slavery, and the South, who wanted to retain it, to put the issues at their simplest. America had to wait until the presidency of Abraham Lincoln before slavery was abolished at the end of the Civil War, only a few decades after it had been abolished in Britain[2]. The pamphlets called *Rights of Man, Part 1* (1791) and *Part 2* (1792) by Thomas Paine, written many years before their time for enactment had come, embraced these same ideals of republican freedom from despotic tyranny[3]. Lincoln was assassinated on 14th April – Good Friday of that year – 1865, less than a week after General Robert E. Lee surrendered his Confederate troops and the Southern cause to General Ulysses S. Grant. Lincoln was born on the 12th of February 1809. Auspiciously, on that same day across the Atlantic, the man whose name is indelibly linked to establishing new ideas of the origin of humankind, Charles Darwin, was born in Shrewsbury, England.

Thus Thomas Paine was a key figure in the humanist social revolutions of the 18th century in America and France that represented the triumph of personal liberty and equality and freedom from authoritarian rule of church or state. Though he did not live to see it, Paine was also instrumental through his writings in achieving the abolition of slavery in Britain and America during the 19th century[4]. After publication of the *Rights of Man*, in the following year, 1793, Paine produced *The Age of Reason* which turned its attention away from social revolution more towards individual religious freedom of thought. An equally influential publication appeared in the same year as the American Declaration of Independence, 1776. That was Adam Smith's book *The Wealth of Nations*, but we shall postpone discussion of that until the next chapter on economics.

Individual motivation

We have seen in the earlier chapters that the quest for happiness is the prime human motivation. A group of social philosophers who sought to disseminate this idea in the 19th century were known as Utilitarians. John Stuart Mill (1806-1873) was a philosopher and sociologist, and godfather of philosopher, mathematician and pacifist Bertrand Russell (1872-1969). He was a philosophical successor to Jeremy Bentham (1748-1832), who advocated a rather hedonistic philosophy. Bentham equated happiness with pleasure and he believed that it was the purpose of every individual and governing system to do everything in their power to maximize pleasure and reduce pain. From this idea, Bentham went on to define a morality of judging actions to be right or wrong depending on whether the effects of those actions promoted or lessened the happiness of individuals or societies: ‘it is the greatest happiness of the greatest number that is the measure of right and wrong’ [23, p.25]. This is the philosophy of utilitarianism. Bentham, J. S. Mill and his father James Mill (1773-1836) rejected the contention of William Paley (1743-1805), a prominent English cleric of the day, that the only motivations for human behaviour were the promise of heavenly bliss or fear of damnation in hellfire[5]. The Utilitarians believed that the more immediate prospects of pleasure or pain provided far greater motivation for human behaviour than some possible future state of bliss or damnation, the very existence of which was far from proven.

Superficially, utilitarianism seems laudable, but it is not difficult to find examples where the increase of one or several person’s pleasure is at the expense of another’s. Obvious examples would be a group of sadists watching someone being tortured, or Romans enjoying the lions feasting on Christians. In the case of jobs in today’s world, when one applicant is successful there are usually a number who are disappointed. When we select a life partner, there are usually a number of competing suitors. So the principle of utilitarianism needs qualification. We can say that the individual’s right to seek pleasure must not be to another’s detriment, as far as possible, though we have to accept that competition is a fact of life, especially in the Western world. Applying this same philosophy, in the present age, there is no excuse for allowing people to smoke in any public arena since it has been established beyond a reasonable doubt that passive smoking by other non-smokers will still be detrimental to their health – possibly even fatally so. The principle of holism demands that we try to find a route to maximum satisfaction for as many people as possible, aware that we will never please all of the people all of the time.

If the human race is to progress, materially and spiritually, there must be a balance between the expression of individuality and the wishes of a community. The personal and the sociological are intimately interdependent. If someone becomes unemployed it is a personal tragedy, but if too many are involved then there are also significant economic and sociological consequences, as we have seen in Britain and America in recent decades with the closure of coal mines, steel mills, car manufacturing plants, and other industrial complexes. As individuals we form part of a social structure. There are always competing objectives that must be considered as to the rights to freedom of the individual against the course of events that is in the best interests of society as a whole, taking into account so far as we can the different preferences of the individuals in that society. As J.S. Mill said: 'different persons should be allowed to lead different lives. . . whatever crushes individuality is despotism, by whatever name it may be called, and whether it professes to be enforcing the will of God or the injunctions of men' [257, Chap. 3]. Mill also believed that men should be free to choose which of the faiths they wished to follow: 'different persons also require different conditions for their spiritual development' [*ibid.*, Chap.3].

Societal motivation

Thus Hobbes, Paine, Mill, father and son, and Locke laid the groundwork in their writings for the evolution of a sense of society among humankind and urged freedom from authoritarian rule by Church or State. Others who followed carried this concept further. The discoveries of science had established natural laws under which the universe seemed to function. A near contemporary of J.S. Mill, the French philosopher Auguste Comte (1798-1857), wanted to explore if there were comparable laws under which a society functioned. He believed that such laws could be determined by a system of 'logical positivism' or 'positive philosophy'. The study of these laws comprised a subject of its own that he called 'sociology'. The positive philosophy was to derive the laws of society based only on empirical observations, as in science. But science always seeks unifying laws that can be expressed in the symbolic logic of mathematics. Today, we have mathematical expressions to represent facets of human behaviour, like sociology and economics, just as we have other kinds of mathematics to express the behaviour of planets and of molecules, and others still to describe subatomic particles. But all our expressions and equations are part of the same single universal field of mathematics, that abstract symbolic logic that the human mind has created to express underlying cosmic truths.

Comte thought that our understanding of the world and the development of society had evolved through three stages. In the first or theological stage humankind interpreted all events as an expression of God's will. In the second or metaphysical stage, represented by the Renaissance, the workings of the world were seen as an expression of human nature, operating under the influence of natural but unknown and untestable forces. The third stage was the positive or scientific stage where the characteristics of society could be attributed to natural or physical laws established by observation and rational interpretation. In 1822 Comte published his first work, *A Plan for the Scientific Works Necessary to Reorganise Society*; his principal work, *System of Positive Polity*, was published in four volumes between 1851 and 1854. In these, inspired by the humanist principles of the French Revolution, he repudiated the precepts of Christianity as a basis for the operation of society and suggested that they be replaced by the principles of reason. Reason is telling us now that the continual pursuit of selfish consumerism is not sociologically or environmentally sustainable.

The co-operative society

Karl Marx (1818-1883), a German-born political economist, had rather different ideas. He saw society as comprising two strata: the capitalists, the moneyed elite who supplied capital for industry, and the working class, or proletariat as Marx called them, who had to seek employment.

Marx also described them as the wage-labour class. The laws governing the evolution of society according to Marx were defined by the class struggle between these two groups. Part I of the *Manifesto of the Communist Party* begins: 'The history of all hitherto existing society is the history of class struggles' [105, p.11]. Marx believed that it was not ideas or values that shaped human society but economic influences. Marx adapted Hegel's dialectical idealism theory of history, shaped by *Geist*, into a dialectical materialist theory of history, where a new class emerges to displace a preceding one as a result of their discontent or alienation. The idea of alienation for Marx implied a loss of feeling of self-worth in that people were valued only for the goods they could produce through their work. Instead of producing goods for their own use as craftsmen they produced components of items that were sold for profit to others. They were subject to forces they could not control — not natural or supernatural forces but forces created by the capitalist employers. They became alienated from one another as property to be bought and sold for their labour. The philosophy of capitalism was that workers should be paid no more than absolutely necessary for their subsistence so that capital would be available for materialist expansion and greater profits for the ruling class. Eventually the working class or proletariat would rise up and overthrow the moneyed ruling class. The egalitarian utopia envisaged by Marx may never be attainable, given the varieties of human nature, but as an objective it is considerably more holistic than the ever widening gulf between rich and poor we are creating today.

The preface to *A Contribution to the Critique of Political Economy* (1859) says: 'It is not the consciousness of men that determines their existence, but, on the contrary, their social existence determines their consciousness'. In Hegelian terms, the tenor of a society determines its *Geist* and that of the individuals in it. Through its materialism, the consumerist society of the West today is more Marxist in principle than the Soviet Union ever was. Marx would not have approved of a totalitarian state that tried to control people's thoughts and ideas and suppress their very essence as human beings. He urged that people should respect the ideal of the brotherhood of man: 'From each according to his ability; to each according to his needs' [241]. Marx's egalitarian ideals are laudable, but the idea that a society prospers through continued material development and acquisition was a reflection of the euphoria of the time derived from the successes of the Industrial Revolution. Today it is the materialism of our society that determines our consciousness. That is why it is so important that we replace that materialist world-view with one of universal spirituality to shape the consciousness of every individual towards sustainable holistic living.

Social identity

Marx's successor in the historical evolution of sociological ideas was the French writer Emile Durkheim (1858-1917), born of a rabbinical family in Alsace. Durkheim applied the empirical methods of science to sociology, looking for statistical correlations underpinned by logical relationships between societal groups. Social facts concerning traditions, customs and beliefs were assessed dispassionately by the observer gathering statistics but they had to be mindful that such practices possessed emotional and spiritual value to the individuals in that social group. Like Hegel, Durkheim believed that such social facts represented a 'collective consciousness' (Hegel's *Geist*) and exerted a coercive influence on the individuals. To some extent, people like to follow fashions in the society in which they live while retaining the freedom to express their individuality. Also like Hegel, Durkheim believed that the social conditions of a cohesive societal group bound them together and had a reality and momentum of their own beyond the lives and perceptions of the individuals in them.

Both Comte and Durkheim subscribed to the sociological theory of functionalism – that a society is an organic whole of interacting components, both practical and spiritual. In the rapidly developing industrial age Durkheim defined two types of social grouping. There were those held together by a 'mechanical solidarity' with respect to the nature of their occupations, while there was an 'organic solidarity' between those with similar ethical values [86]. In either case, individuals defined or reinforced their personal identity from the social group with which they associated. As we have seen in previous chapters, this kind of social integration is considered by psychologists to be an essential component of fulfilment and holistic living. Durkheim followed Hegel in believing that those who could not conform to the beliefs or practices of a society or social group felt a sense of alienation from their social environment. Durkheim published a seminal work on the subject of suicide, which he saw as an extreme tragic consequence of an individual's alienation from society [87].

Each social group expresses its personality through its own traditions, as we saw in Chapter 4. These are influenced greatly by the predominant religion, which determines life choices for the individual, like dress and eating habits. Within this cultural identity there is now also increasing cultural diversity in many countries through migration, though ethnic groups tend to live within cultural ghettos in a foreign land. If we are to achieve social cohesion then those who migrate from one country to another have to be prepared to embrace the culture of the land into which they move to a large extent. They should not have to abandon their religion or cultural practices, but if they insist on retaining all of their former cultural traditions, especially those that are culturally divisive or offensive to the native citizens, and reject those of their adopted land, we will produce a very fragmented society that lacks any kind of cohesion or national identity and holds within it the potential for explosive cultural conflict at any time. If migrants are adamant about retaining all such divisive traditions, then they should stay within their native culture where such traditions are respected.

Each of those individuals who are part of a coherent social group has their own objective social identity that reflects the ways in which they are the same as others – through their ethnicity or religion or profession or membership of some political or sporting club. But they also have a subjective personal identity that sets them apart from others – to use Jungian terms, the persona they present to the world and the self that reflects their inner being. Psychologists Adler, Fromm and Horney came to the same conclusion as Comte and Durkheim from the sociological standpoint – that individuals develop this sense of self through their interaction with others [94].

The evolution of society

Like Karl Marx, Max Weber (1864-1920) was born in Germany. He was a sociologist, economist, historian and philosopher. Although accepting Marx's view that economic factors were important in producing social change, Weber believed that ideas, values and beliefs – reflected in Hegel's *Geist* – were the principal factors motivating individuals to act in concert to bring about change. His specialism was a study of the interaction of religion, sociology and economics. Weber studied the religious philosophies of East and West and concluded that the focus on the individual in Christianity was a significant factor contributing to the rise of capitalism [380]. The organized bureaucracy of large companies he said was an application in the world of business of the rationalization of science. But where the physical sciences demanded dispassionate and objective rationalization, social science situations needed sympathetic understanding of people's values, which was absent in large corporations.

Weber's definition of sociology was that it was 'a science concerning itself with the interpretive understanding of social action and thereby with a causal explanation of its course and consequences' [381, p.4]. If values and beliefs are the most important factors in producing social change, then there is an urgent need to inculcate into the unconscious of society a belief in the need to live in harmony with one another and with our natural environment – to disseminate a holistic spiritual Nature religion.

The most important function of religion surely is to teach us how to behave in relation to others – to teach us how to live fulfilling lives for ourselves while implementing the Golden Rule of the ancients or, in more recent philosophical terms, to live Immanuel Kant's Categorical Imperative: 'Act only on that maxim through which you can at the same time will that it should become a universal law' [198, section 30]. English philosopher Thomas Hobbes was of the same opinion: 'every man that hath attained to the use of reason is supposed to know he ought not to do to another what he would not have done to himself' [172, Chap. XXVII]. But passively not harming others is not enough: as Eastern philosophy teaches, and some of the Western religions have followed, we must work continually to improve the quality of life of others in a positive way.

In Christianity, which has formed the basis of the development of western civilization, the emphasis has been more on personal salvation than social cohesion or environmental sustainability. In this sense it has been a divisive force, taking us away from social unity: Judaism, Islam, and even the atheistic Confucianism are far more cohesive of family and society. Furthermore, the fundamental role of Christian missionaries was to destroy holistic and socially cohesive pagan Nature spirituality and replace it with the synthetic edicts of the Catholic Church. As J.S. Mill said: 'What little recognition the idea of obligation to the public obtains in modern morality is derived from Greek and Roman sources, not from Christian. . . I believe that other ethics than any which can be evolved from exclusively Christian sources, must exist side by side with Christian ethics to produce the moral regeneration of mankind . . . in an imperfect state of the human mind the interests of truth require a diversity of opinions'. Mill saw Christianity as a 'grave practical evil' and maintained that 'a large portion of the noblest and most valuable moral teaching has been the work, not only of men who did not know, but of men who knew and rejected, the Christian faith' [257, Chap. 2]. One of the themes of this book is that, amongst what we may call 'organized religions', the eastern religious philosophies offer more socially cohesive holistic ideas than those of the West.

Sometimes we despair at what seem to be setbacks in the spiritual evolution of humankind when we read of continual civil wars, authoritarian dictatorships and religious persecutions. But slowly, agonizingly slowly and painfully, humankind does seem to be moving in a general direction of greater tolerance and spirituality. Hegel believed that history has a purpose and, for those with a holistic world-view, that purpose is evolution of society to a state of greater spirituality through *Geist*, regaining that reverence, awe and wonder for the natural world that characterized the lives of indigenous peoples.

This tortuous pattern of the evolution of the human psyche has been described by some contemporary psychologists as Spiral Dynamics[6]. Certainly human evolution, whether biological or social, has not progressed thus far in a smooth continuous upward line; it proceeds erratically with many lateral transitions. Further back along our evolutionary path, many millions of years before the first hominids, there were several massive extinctions of various life forms – at least five times – and some estimates say that 99% of all the life forms that have ever lived on Earth are now extinct. Our urgent goal now is to develop a spiritual mind-set to see that we do not precipitate the sixth extinction!

So even before the appearance of the most destructive species on Earth – Man – Nature itself has come close to extinguishing life on Earth. As the Anthropic Principle suggests [20; 47], Nature had to establish an exacting set of criteria for the creation and evolution of life in the first place. We would therefore perhaps be too optimistic if we expected that human social evolution would proceed smoothly to harmonious universal spirituality. With natural disasters periodically taking their toll of human life we would do well to avoid killing each other, even if world population is already far greater than the natural world can sustain. The way to stabilize it is by control of population growth at its source – conception – not to have people dying from hunger, disease, war and exterminations in 'ethnic cleansings' because Earth's resources are simply not sufficient to maintain our current materialist profligacy.

In the Medieval period we had bouts of plague to take their toll on human life. Even in the early part of the 20th century, with all the advances in sanitation and the emerging understanding of the principles of nutrition, still a pandemic of influenza killed a quarter of a million people in Britain and twice that many in America. In 2003, 100,000 people who were admitted to hospital in Britain picked up an infection there and 5000 of those died from it [177]. We still have huge ongoing battles to fight against disease, the misuse of drugs and the (over)use of limited environmental resources. These challenges call for international cooperation and a holistic outlook amongst all participating parties at every level of society. Humankind cannot afford to indulge the egos of dictators, whether political or religious, and disseminating an all-embracing societal spirituality is an urgent priority.

Anyone with any sort of social awareness cannot fail to have noticed that a shift in the global consciousness *is* taking place towards greater integration. There is a greater interest in mysticism and psychic phenomena, and people are turning once again to traditional therapeutic medicines and treatments, partly from disenchantment with diagnoses and treatments from orthodox medical practitioners and partly from unpleasant experiences of the side-effects from allopathic drugs. But it also represents a return to a more natural and less artificial, materialistic philosophy.

There are many shamans and sages around the world who talk of this shift in human consciousness to holism [238; 308; 321; 394]. Over the past two millennia there have been many predictions of either the Second Coming or global catastrophes and Armageddon on specific dates, and many contemporary seers are talking about the year 2012 as being in some way critical [33, p.53; 359]. This date has been found also in several ancient charts of divination, from the Mayans and Aztecs for example. The pessimists foretell a global disaster, such as could be caused by a natural event like a huge volcanic eruption, or by man-made global war. The optimists however say that it could be a time when the world finally wakes up to the need to work together in harmony, conserve our natural resources and stop plundering and desecrating our Earth environment. Perhaps there will be an international treaty that the most powerful nations on Earth not only sign up to but actually implement – immediately! As we saw earlier, the Swedish polymath Emanuel Swedenborg (1688-1772) believed that the Second Coming of Jesus on Earth was not a physical reincarnation but a realization amongst humankind of the need for this kind of global harmony, inspired by the example of the life of Jesus [355]. So persuasive were his arguments that his followers founded the Church of the New Jerusalem in America around 1784 to extol and disseminate his principles.

We are reminded of the vision of human evolution suggested by the French mystical palaeontologist Pierre Teilhard de Chardin (1881-1955). He contended that the world has passed through a series of evolutions, both material and spiritual, but that the focus now was on the spiritual. First there was geogenesis, the creation of the land and the sea, the physical shaping of the Earth and the drift of the continents to their present positions. Then came biogenesis, the creation of life from the simplest of single-celled organisms to the complexity of humankind. This was followed by psychogenesis, the development of thinking beings; we know that early man had a smaller capacity brain but that this grew in size as Man evolved. Now we are in the stage of noogenesis, the evolution of mind into soul and into Communal Soul [364, p.201]. This is a more sophisticated version of the Medieval vision of a Great Chain of Being – the four 'kingdoms' of Creation in mineral, plant, animal and human expressing matter, life, consciousness and awareness, with what we would now describe as increasing sensitivity to the universal spiritual field as one ascends the chain. The distinctive quality of the human organism is this capacity of self-awareness and empathy with others in a way that far exceeds any genetically programmed herd instinct that is found in animals. We also have the capacity to commune with a spiritual dimension greater than ourselves that is evident also in some other animals [334]. The question now is how we use this faculty – do we simply ignore what reason is telling us and continue in our own selfish, hedonistic way or do we heed our intuitive input and take action to prevent potential environmental catastrophe?

To all non-industrial people, human existence is inseparable from the natural world and from the universal spiritual energy. The concept of belonging to some limitless unseen seems to have been a feature of life even in human pre-history. This has nothing to do with religion, as such; it is simply a feature of human existence. It is a characteristic of human behaviour to want to feel a part of something larger, whether it is a concept of universal spirituality or, more trivially, to be a member of some club or society. The sense of belonging is important in itself, and we define ourselves by the relationship we have with others. This human characteristic of feeling intimately bound up with the whole world outside of the body is particularly prevalent in less technologically developed indigenous societies. The very fact that these less-developed peoples live in dwellings made of natural materials gives them an intrinsic awareness of contact with the spirit of the Earth.

The lessons for modern society

Religion generated such antipathy from some social reformers like Hobbes and Paine because it placed restrictions on individual freedom by telling adherents what and how to think, and even how to behave regarding matters that did not concern society at large, supposedly for their individual spiritual salvation. This is particularly the case in Western religion, where scripture, and its interpretation according to the religious hierarchy, is at the heart of the faith; it is less so in the East, where individuals are encouraged to find their own route to salvation, and the hierarchy guide rather than instruct. Religion is a social activity created by humankind (whatever the scriptural source claiming divine revelation may be) to establish and reinforce principles of individual and social behaviour for the stability of society. Some religions, particularly Judaism, Islam, Baha'i and Confucianism, emphasize this social aspect –

to serve the community through one's individual practice, to honour family bonds and to establish social bonds through one's conduct and the education of others by instruction and by example.

The human spirit is such that it cannot find full expression under a system of repression, whether by Church or state, but the expression of individuality must be directed towards a common good if we hold a holistic philosophy. Standards of discipline in schools and in society at large have declined in recent decades with the advocacy of 'free expression', and the 'rights' of the individual, such that one person's right to satisfaction so often leads to another's impoverishment. One disruptive and undisciplined pupil can destroy the cohesion and capacity to learn of a whole class. One person's freedom to drink alcohol or smoke or take other destructive `recreational' drugs when or where they choose can have appalling and tragic consequences on the lives of so many others. Without paralyzing legislation, the only solution to these social problems is for each individual to take responsibility for their own behaviour to see that it is not detrimental to others.

Ours is an age that is preoccupied with continual material growth. Social status – being a part of the 'in-crowd' – depends largely on material wealth and power, now, as it has done in the past. The difference today is that, with the technological developments since the Industrial and Electronic Revolutions, we now have a greater variety of sophisticated tools and toys we can possess as a sign of our status – fast cars, a private aeroplane, a yacht, two or more grand houses filled with electronic gadgetry, and so on. And with the increased possibilities for acquisition of such wealth, the gap between rich and poor ever widens, for both individuals and nations. Those that have the greater wealth and power also have a greater responsibility to use it wisely. The developments of the Industrial and Electronic Revolutions have given us an enormous and exciting potential for human fulfilment, but with that capability for expansion of human horizons has also come greater power for destruction. We need to apply what the American sociologist C. Wright

Mills calls 'sociological imagination' in how we use this expanding knowledge – always keeping sociological implications in mind in what we do [260].

One of two major themes of Mary Midgley's book *Evolution as a Religion* [253] is that the 'survival of the fittest' self-interest interpretation of Darwin's theory of evolution is in error and that humankind has evolved and continues to prosper more through cooperation rather than competition. While I endorse the sentiment, I think the competition motive reigns supreme in our capitalist society today and its influence is far more pervasive than cooperation, however desirable the latter may be. Perhaps until the 20th century, even until the latter half of the last century, we did progress primarily through cooperation, but the need for 'accountability' and the unceasing quest for profitability as the *raison d'être* of any enterprise has made cooperation subservient to ruthless competition.

Man is innately a social animal. We have seen in this and earlier chapters the part that social interaction plays in the development of the individual. It is part of his genetic make-up to exhibit what Kant called *humanitas*, or 'a habit of harmony with all other men' [197, p.232], which gives us our words 'humanism' and 'humanitarian' – a concern for other people. This quality was recognised by the ancient Greeks and they described it as *koinonia*. In his *Politics* [III, vi, 1278b], Aristotle declares that 'men have a desire for life together even when they have no need to seek each other's help.' The English metaphysical poet John Donne (1572-1631) wrote 'No man is an island, entire of itself: every man is a piece of the continent . . . any man's death diminishes me, because I am involved in Mankind' [82]. It is this mutual concern that binds humankind together. It is what David Selbourne calls the 'civic bond': 'The civic bond rests, in the first instance, not upon shared rights but upon reciprocal obligations' [327, p.78]; and further, 'the health of the civic bond is a prerequisite for the well-being of individual members of the civic order' [327, p.25]. As an Editorial in the British broadsheet, *The Independent*, commented a few years ago: 'The values of community, duty and responsibility can only return when the British regain what De Tocqueville[7] called the "habit of association", some sense of loyalty and commitment to things that were lesser than the state but greater than the individual or the immediate family' [347].

There is an upper limit to the size of a coherent community. It is well known that those who live in villages usually know what occurs in the life of all the other villagers – for good or ill! Sometimes, this may feel like intrusion into our privacy but, on the other hand, such mutual concern can be of vital importance if we are ill, or have undergone some trauma, or if we simply need the help of our neighbour in some mundane task. In the large city, on the other hand, people tend to live more secluded lives. The opportunities for social gatherings may be much greater, but such acquaintances tend to be fleeting and rather superficial. The incidence of crime also tends to be much greater in the large city than in the village because of this state of anonymity and lack of social cohesion.

The same principle applies in the corporate domain. Small companies value the contribution of each employee as an individual. In large companies, employees are no more than entries in the payroll statistics. With mergers and acquisitions, expanded global corporations are becoming ever more impersonal and ruthless in their pursuit of profit. The welfare of staff is of secondary importance. Schumacher made much of the dehumanizing aspects of large corporations in his book *Small Is Beautiful* [324 – see Chapter 6]. In the wake of Descartes and the success of scientific materialism, during the latter half of the 20th century organizations were structured and operated like machines. This philosophy was highly successful in increasing productivity and overall efficiency, but it also generated a great deal of stress for managers and workforce alike. There is a trend that is just beginning to develop for organizations to view themselves as living social communities in which employees can realize their personal goals along with those of the company. Taking his lead from a study by Arie de Geus [79], Fritjof Capra says: 'a strong feeling among the employees of a company that they belong to the organization and identify with its achievements – in other words, a strong sense of community – is essential for the survival of companies in today's turbulent business environment' [45, p.95].

The criterion of optimum size applies also to schools. The philosophy of creating schools with ever increasing numbers of pupils is that such a centre can offer more opportunities for study in a range of subjects with the best of facilities. The drawback of this – and it is a huge one – is that teachers do not get to know pupils as individuals, with their personal needs, strengths and weaknesses. For the pupils themselves, there is a much greater sense of security, of community, of camaraderie in a school of a few hundred pupils compared with those where pupil numbers exceed a thousand (or several thousand in some schools in large American cities), which many pupils find intimidating, especially in the early years. Discipline then breaks down as pupils no longer feel a sense of belonging to the school community or responsibility to others: disruptive behaviour is often no more than attention-seeking by a pupil who feels neglected. The environmental consequences of having to 'bus' these large numbers of children considerable distances are as wholly negative as having adults commute great distances to work. A return to smaller local schools would, overall, be beneficial.

A major problem in Western society, and particularly in Britain and America, is the lack of respect that children have for authority figures like parents and teachers, arising from a lack of their own self-respect and a feeling of helplessness in the face of overwhelming factors. The same is true for citizens in general — a lack of respect for the politicians whom they have elected to govern them and for the rule of law because of this feeling of insecurity and helplessness when faced with the increasing complexity and bureaucracy of modern life. This disrespect and social antagonism is fostered and encouraged by behaviour exhibited on television, which now seems to command several hours of viewing time in most homes in the West. It is debatable whether television is the largely beneficial educational medium that some educationalists claim. Certainly, there are many fine educational programmes, but their number is far exceeded by programmes for popular entertainment, many of them displaying aberrant social behaviour. The argumentative and aggressive behaviour continually displayed in 'soap operas' indoctrinates viewers, especially impressionable youngsters, with the idea that this is the normal and desirable way to behave and to resolve differences of opinion. If this really is normal family behaviour, then we have a very long way indeed to go to establish a holistic outlook in society. It would be far more socially advantageous and would be no less entertaining to see stories of constructive social interaction and achievement that produces positive outcomes in overcoming challenges rather than focusing almost exclusively on dysfunctional relationships and family breakdown [see ref. 338].

Television producers would have us believe that the medium only follows societal trends and does nothing to change them. If that were so, commercial organizations would hardly spend millions of pounds or dollars on advertising if they were not going to be rewarded with increased sales by brainwashing viewers with the same repeated messages that inveigle their way into our unconscious. As Sue Palmer remarks: 'The messages that marketing feeds daily to our children now amount to a gradual . . . subversion of civilised values' [285, p.239]. They send a message of endorsement if not encouragement for children to break rules set down by parents and teachers: 'big business now engages in complex mind games with children' [285, p.233] to the detriment of our children and ultimately to ourselves. Susan Linn comments: 'the explosion of marketing aimed at children today is precisely targeted, refined by scientific method and honed by child psychologists – in short, it is more pervasive and intrusive than ever' [220]. As the American marketing strategist, James McNeal, openly claimed: 'Kids are the most unsophisticated of all consumers; they have the least and therefore want the most. Consequently, they are in a perfect position to be taken' [247].

Capitalist business, promoted by advertising executives, has one motive and one only – profit. We must never allow ourselves to think that they ever have our best interests at heart. Greed is the only motivator of capitalism and morality is not an issue. We cannot sit idly by and just watch this happen. We must organize ourselves to fight this insidious brainwashing and exploitation. This type of social change will only happen if more individuals in the electorate and thence politicians adopt a holistic world-view embracing a spiritual sense of community. The encouragement to buy more and more nationally produced goods whether or not they are really needed, as if we were being patriotic by so doing, is misplaced. If goods are essential, then certainly buying local products is more holistic and ecologically sound; but distribution of goods originating even from other parts of the same country involves a negative ecological effect.

The adverse effects of poor diet and insufficient exercise on children's health and behaviour were referred to in Chapter 1. This is a social issue. Playing computer games can be fun and educationally stimulating. But too many hours in front of a computer screen poses potential health risks from the electromagnetic radiation [274] and takes time away from equally enjoyable but much healthier outdoor or social activities [228]. Too much time sitting passively before a computer or television set is likely to stunt physical and mental development, however much educational benefit may be gleaned. Reading forces the mind to convert black and white squiggles on a page into imaginative mental images: on a TV set or computer, the pictures are already there. Television sets and computers in study bedrooms also tempt youngsters into spending too much time on what they regard as play and too little time in boring sleep. But that sleep is where the images of the day, provided there are enough of the right ones, get consolidated and the young mind prepares for the world of tomorrow. Between 2000 and 2002, 68% more mind-altering drugs were prescribed to children in the UK alone [285, p.17]. This is an appalling reflection of the state of young people in our society. It is time that governments, education authorities and the medical profession got together to address the root causes of this situation instead of placating disturbed children with drugs. At a conference in 2006 at the Centre for Social Justice, Brighton psychologist Melanie Gill maintained that the problems of disaffected and rootless adolescents stem from the lack of respect for the family as the fundamental unit of society and with it the undervaluing of the role of motherhood [123]. Too many mothers today choose or are forced to go out to work: but motherhood is not a part-time hobby and must be valued more highly if we are not to disintegrate as a society.

The regrettable breakdown of nuclear family groups and the almost total disappearance of extended families of blood relatives means that children often no longer benefit from input from the lives of grand-parents, aunts and uncles, who can also ease the strain on harassed parents occasionally with child-minding. This situation also arises because people have to move away to find work or are involved in extended travelling in their jobs. The effect on children is likely to be almost wholly detrimental, even if it does teach them a degree of self-sufficiency. Too often children are placated with material gifts to keep them occupied while overworked parents try to cope with a job, housekeeping and child rearing. This encourages in children the idea of instant pleasure and does nothing to inculcate the ideas of self-discipline and deferred gratification – that hard work and effort may be required to produce greater rewards at a later date. What children really need is quality time for love and guided human attention. We need some social organizations where harassed parents can get practical and emotional support. Today, it cannot be only for mothers with toddlers as fathers are playing an increasing role in family life. It would also be more accessible and effective if it were non-denominational with no attachment to any religious group.

Even when both parents are present, the demands of keeping up a place in society may force both out to work and then a surrogate must fill in with child-minding duties. These social changes may make many luxury items around the home more affordable, but they are highly erosive of family life and of holistic social cohesion, even if they superficially enrich the quality of life for the individuals in the family. The absence of parents to tell fairy tales or adventure stories, or tales of their own exploits or those of other family members, means that children are denied that stimulation to their imagination and getting less contact with their own heritage, which leaves them in a cultural vacuum. Without family supervision, children cannot be guided into learning to develop extended concentration. The continually changing images on TV programmes and videos encourage this idea of constant change and lack of focus. While I was school-teaching, we were told at a training day that children should be encouraged to study for about 20 minutes at a time, and then take a 10 minute break before resuming. When I was a student myself, we were told that it takes 20-30 minutes to build up focused concentration and that study should then continue for 2-3 hours (a whole morning or afternoon, though not necessarily on the same topic) before taking an extended break, preferably for a meal. The youthful brain cannot have evolved to such a different state in just fifty years unless it is by indoctrination from the constantly changing images on television, video games, and games in amusement arcades. Learning more extended concentration is part of the process of developing self-control as well as developing the neural connections in the young mind. Mark Griffiths has pointed out that the dangers of producing addiction to arcade games in teenagers is as great as that for drugs, and just as dangerous in a different way [137, 138]. Much of the reason for the lack of respect that children show towards their elders and authority figures is a reflection of the lack of respect they have for themselves.

The foregoing arguments make a case for the belief that individual fulfilment demands service to the family and society as a whole. This is the principle of duty, so eloquently expressed by David Selbourne in his book [327]. This presentation echoes J.S. Mill's thoughts on the subject a century earlier. Mill wrote: 'every one who receives the protection of society owes a return for the benefit and the fact of living in society renders it indispensable that each should be bound to observe a certain line of conduct towards the rest' [257, Chap.4]. Mill tells us that 'no person is an entirely isolated being' and John Donne in *Meditation XVII* says that 'No man is an island'.

It is this sense of duty, of belonging that has been lost from western society in the focus on self-interested pursuits and the assertion of the rights and freedoms of the individual: 'while duties without rights make men slaves, rights without duties make men strangers' [327, p.108]. Increasing urbanization tends to lead to increasing isolation rather than to a greater sense of community. Selbourne follows Adler, Fromm and Jung in pointing out that every human being is a part of a social enterprise and that achievement of fulfilment and well-being for each individual person – Jung's individuation – is dependent in large measure on their relationship with society as a whole. Even in the 19th century Mill felt it necessary to urge: 'there is need of a great increase of disinterested exertion to promote the good of others' [257, Chap.4]. The bond between individual citizens, or between a citizen and some specialist political, religious or other group, is more important, more intimate than the relation between a citizen and some more impersonal concept such as the state, as those countries who had adopted a Communist philosophy found in the late 20th century.

Society is the very fabric woven from the strands of individual lives. The whole *does* have an identity of its own, and it has the potential to be greater than the sum of its parts. All individuals cannot prosper if the society that surrounds them is spiritually or materially impoverished. It is arrant nonsense to say that there is no such thing as a society, only a collection of individuals [365]. But it accurately reflects the tenor of the age in Britain over the past few decades and in America over the last half-century when the Nietzschean cult of an individual's will-to-power has been presented as the desirable goal for all. But it was, in the main, only those who already had power who were able to further exploit it at the expense of others, increasing the fragmentation of society. The introduction of obscene levels of prize money in a national lottery is but another symptom of the divisiveness of the age — making some people absurdly rich, for no expenditure of effort, while others work diligently for a pittance all their lives or live in cardboard boxes on the streets. Such a national gambling set-up cannot fail to breed resentment and alienation as well as encouraging another human weakness that all too easily becomes an addiction.

Organized gambling should be intolerable to any civilized, caring society. There must be more equitable ways of distributing charity to worthy causes. If the lottery cannot be abolished, simply reducing the size of the prize money and dispersing it to more people would be a step in the right direction. It is particularly important that adolescents should not have the opportunity to gamble for, as psychologist Mark Griffiths[8] has pointed out, it can soon become addictive and self-destructive. The introduction of 'super casinos', as is now occurring in Britain, is another step in the wrong direction, away from holism. Since the introduction of such casinos in Australia, the number of those addicted to gambling has increased according to Professor Mark Dickerson of the Australian Institute of Gambling Research such that some 1-3% of the population now have a gambling problem. Such an addiction affects not only the life of the gambler but that of many other people with whom he or she may come into social contact.

Mere aggregations of people do not make a society – the active participation of each individual, so far as they are able, in building communication with others is an essential component of any stable society. This sense of civic duty arises, in the first instance, from parental instruction and example; then the responsibility falls onto professional educators, thence to individual adults taking responsibility for establishing bonds with others. Parents, schools and community are the sources of social stability and holism. Where one fails, the others should be willing and able to help shoulder some additional responsibility, but this does not mean that parents can abdicate their parental responsibilities for their children and leave all matters of instruction and discipline to school teachers and issues of care and nurture to the state.

The sense of duty is something that needs to be inculcated from the earliest days of infancy for the infant will sense, if not cognitively understand, the devotion of others to their welfare. It would be encouraging to suggest that the display of such concern for the human and natural environment is innate, but pressures of modern urban life seem to create so many distractions from what was once – and still is amongst indigenous peoples – an automatic part of the fabric of life. Such dedication to others needs to be voluntary and heart-felt – hence the need for infants and children to be surrounded with love, care and concern. As Cicero says: 'an action, although honest, is not truly virtuous unless it is done out of choice and with a good will' [53, I, ix]. The principle of duty derives from ethics, borne of faith, and from reason and utility in serving self-interest through helping others. It may be said, to use Jungian terminology, that the disposition to care for others is an archetype that finds first expression in parental care of offspring and the mutual concern between siblings. In western society, rather more than in the East, parental responsibility for children is too often sacrificed in the need or desire for material acquisition, while care for the aged has often been transferred to the state as extended families living together or in close proximity are becoming increasingly rare. This last situation is to some extent necessitated by the fact that people now live longer and tend to have a greater number of health problems that require professional care, but it is also a consequence of the expectation that, in the capitalist economic system, professional workers must be prepared to travel extensively to or during their work leaving less time and energy for family responsibilities.

No society will ever be truly egalitarian – that is an unrealizable utopian ideal. There will always be energetic and enthusiastic individuals and lazy ones; those who are endowed with special gifts of ability and those who are not; those whose focus is on self and those who devote themselves to the welfare of others. If we are to encourage individuality of expression and those with imaginative or entrepreneurial skills, it is inevitable and right that this should be so. We do not want all individuals to be treated the same: as Karl Marx said: 'from each according to his ability, to each according to his needs' [241]. What a just society requires is that there should be equality of opportunity, appropriate to the needs of the individual, though this attitude has become unfashionable in the current materialist social climate. The very formation of a society means that the freedoms of the individual will be curtailed so that they do not impinge on those of other people. A just and contented society is one which is best able to find the balance between individual freedom and state control so that the greatest number of people can live largely in peace and contentment. If the people resent government restricting their freedoms, they must accept that they behave responsibly as individuals towards others: if they do not, they will be forced to do so by legislation.

The goal of existence for each individual is the satisfaction of needs. Food and shelter are among the most urgent needs; but these form a part of a greater overriding quest for pleasure and avoidance of pain. This is what psychologists call the 'pleasure principle'. The mark of sophistication and civilization in a society is that its members have the time and inclination to pursue interests that seem to be of no immediate practical use, like art, poetry and music, but yet which will, in the long term, help breed contentment and develop a spectrum of skills in the individual which can often be turned to good effect for the benefit of society. This lesson needs to be learned in childhood by example from parents.

Self-restraint of material demands and hedonistic desires is an essential component of individual and societal wellbeing. Prohibition of alcohol use in the United States in the 1920s was a disaster. The rule of law was universally flouted and the setting up of premises where alcohol could be obtained was instrumental in the establishment of the organized crime network that now influences so much of American society. There are many healthy ways of deriving novel experiences in life without the use of toxic drugs like alcohol, tobacco, cocaine and heroin. These not only debilitate the individual. The consequences of their use ripple on to cause problems in society as a whole – the need for emergency medical treatments, vandalism and other offensive behaviour, and traffic accidents caused by drivers whose judgement is impaired. There is no question that the best interests of society would be served if all such substances were prohibited. However, we cannot uninvent the wheel. These substances are with us now and all we can do is encourage and educate people as to their dangers and their social responsibility so that we minimize the adverse effects on our society that would be caused by irresponsible individuals. As Barbara Brown Taylor has said: 'We belong to a web of creation in which nothing, absolutely nothing, is inconsequential . . . Every one of us will change the world, whether we mean to or not' [361]. The message therefore is to be sure that our intentions and actions are constructive and not destructive. Every decision we make and every relationship we share have the potential to take humankind closer to the expression of holistic spirituality on Earth.

Through our education system, we need to inculcate the values of duty to the community into our children, a whole generation of whom have grown up in this self-obsessed, profit-driven society, while often lacking basic skills of literacy and numeracy that are fundamental to their education and ability to earn a living. The values advocated by sociologists in this chapter are not just religious or ethical values of some theoretical moral high ground, though they would be none the worse for that: they are of immense practical importance in securing the stability and development of the society in which we live and thence, our fulfilment as individuals. As Aristotle remarked in the *Politics* [VIII, I, 1337a]: 'the responsibility for education must be a public one . . . In matters that concern the public, training for them must be the public's concern'.

Rabbi Dame Julia Neuberger has some trenchant comments about the decline of social cohesion and the need to find some sense of moral purpose as a society, whatever our religious beliefs: 'Unless we rethink our social obligations and reassess the value of trust, we will become even more cynical, even more atomistic, even more individualistic – and there really will be no such thing as a society' [269]. These thoughts are reflected by Sue Palmer: 'it's the balance of individual rights with social responsibilities that underpins life in a democratic society' [285, p.297].

We also have given insufficient attention to emancipating the role of women in society. Success in our capitalist society has become associated with establishing aggressive behaviour to outdo our competitors, as we imagine would have been the case with cave men before they learnt that more could be achieved through cooperation than by confrontation. The story of Ladakh is a sobering contemporary example of societal cooperative harmony being disrupted by Western capitalist philosophy into an aggressive and competitive order: where we value possessions, they value community. In fact, native peoples are being displaced and exploited by western capitalism on every continent [236]. The economics and social structure of this tiny Himalayan community is described in full in Chapter 6, but a preview features shortly. Hobbes says: 'The passions of men are commonly more potent than their reason' [172, Chap. XIX]. Baruch Spinoza (1632-77) shared this view: 'Men are led more by blind desire than by reason' [352, Chap. II, 5]. Men more so than women? Our societies for two thousand years and more have been predominantly patriarchal. It has been the role of men to make the important decisions of government, of business, and even of family. Aggressive decision-making has been seen too often as a virtue, a sign of masculinity, and a fitness to lead. Women have been forced very largely into a subservient role, and were valued for their obedience, subservience and 'femininity'. Only now are we beginning to realize how much richer and less confrontational our societies would be if the intuitive grasp of women was allowed to play a more prominent part alongside the aggressive rationality of men. Of course these are the established stereotypical images of the sexes and such qualities are by no means universal, but aggression is unquestionably to be found more often in male-dominated western society than is beneficial.

This focus on the individual, and the enforced abandonment of pagan communion with the natural world, though some may give these developments a patina of authority by stating them as Christian dogma, has contributed to the spiritual decline of society so evident in Britain and the United States. Political systems in both nations in the last decades of the 20th century have exacerbated the situation with their individualistic material philosophy. The gap between rich and poor (individuals and nations) has increased, generating justifiable resentment amongst the dispossessed and putting still more emphasis on material ends and a quest for continual material growth. In an age of specialization, mediocrity is expanding, as those who do not have the opportunities of acquiring the necessary skills fall ever further behind. This is obviously detrimental to social cohesion and holistic growth within or between societies.

The New Story that several contemporary mystical ecologists like Berry, Hartmann and Swimme are advocating needs to embrace some of the social principles that govern societies of indigenous peoples [27; 153; 360]. They have political independence and live in an egalitarian society, knowing all the other members of the group. There are shamans who are respected as elders of the tribe with special powers or possessing wisdom commensurate with their age, but decisions affecting the tribe are taken democratically. They get their resources locally and live within the means that their natural environment provides. They place great store by social harmony and avoid confrontation between individuals in the tribe or with other tribes if possible. Life is lived in harmony with a natural spirituality, which provides contentment and much leisure time, though is often harsh and arduous. How this 'primitive' lifestyle contrasts with our western 'civilization'!

In the West, individuals are slaves — slaves to managers, production targets and profit margins and they continually need to seek favours from those higher up in the hierarchy to make their lives more bearable.

Outside of the workplace, the general parameters of our lives are dictated by government bureaucrats whom we never know or see. The lives of Europeans are governed by a parliament made up predominantly of members from other countries who know little or nothing of those factors that are important locally. The resources that sustain the community are obtained world-wide, with little knowledge of local conditions or effects of trade. If peaceful reciprocal trading fails, there is always war and conquest with spoils for the victor. Life is lived according to the dictates of politics or religion. It is generally physically comfortable for those with assets, but increasingly stressful and providing a decreasing amount of leisure time for most.

Unrestrained, indiscriminate and undiscerning power in one individual, or in one nation, or in one religion cannot fail to produce disharmony. Such power, and the rebellion against such power, often leads to violence as we have seen repeatedly throughout human history and on into the 21st century. Have we learnt nothing from the lessons of history? Must the demagogue and his followers be allowed to destroy our civilization? When the state exerts any restraint on individuals, the liberals always rebel with cries of 'nanny state'. In Britain, such was the response to suggestions of a ban on foxhunting, or prohibition of smoking in public places, although these were humane and healthy policies. But restrictions on individual freedom must be imposed, for one man's expression of freedom can too easily infringe that of another. As Kant said, if there were no restrictions of individual freedoms, there would be 'no certainty that man might not use his powers to destroy himself, his fellows and the whole of nature' [197, p.122].

Our definition of 'progress' seems to be 'change'. In our eagerness to embrace all that is new we have lost sight of all that has been achieved in the past and stupidly make the same mistakes over and over instead of learning from the experience of our predecessors. There is no respect for experience – of parents, of teachers, or of those in authority who should have achieved their positions by merit in an equitable system. Because of the inadequacy and corruption of the system for promotion within organizations, too often it is those whose primary concern is with personal career advancement who reach the higher levels in ruthless ambition so that they can exploit those over whom they have management control and capitalize on their ability and commitment. Too often, people – and usually men – are promoted higher and higher in an organization in the belief that they will then no longer be in a position to *mis*-manage – but still they accumulate all the perquisites that go with their elevated status, including generous compensation if they are dismissed for incompetence. Such organizational practices are incompatible with a successful and holistic society.

Despite the antagonism towards organized religion in some quarters, partly because it represents a return to what is regarded as superstition over rationalism, and partly because the acts of terrorists have made it seem socially destructive rather than cohesive, there is, surely, a role for religion in promoting the ideals of a coherent society. However we envisage the Divine or Ultimate Being, religion is something that could give us a reason for living – and for dying. As human beings, we cannot be complete without that expression of our own spiritual individuality in relation to others. Human wants are infinite, but human needs are few – and a faith to live by, nurtured by a relevant and percipient religion for those who wish it, is one of them and is an integral part of a holistic societal life.

The lesson of Ladakh – a preview

This topic will be dealt with more fully in the next chapter on economics. Here, I want to emphasize some of the social consequences of the changes that have been brought about in this isolated Himalayan community.

The societal decline of the territory of Ladakh presents a salutary lesson for us in how to destroy a coherent society by imposing Western individualistic and materialistic objectives. Helena Norberg-Hodge has spent sixteen years among the people of this once-isolated country and has witnessed the moral and societal decay of this country that has come about primarily by the implementation of wholly inappropriate western economic philosophy. Ladakh is a part of the Kashmiri region of the Himalayas bordering on Tibet and it has come to the attention of those in the West who are concerned with the effects of over exploitation of people and land and the corrupting influence of trying to impose a Western lifestyle with its materialist values on all peoples everywhere. Helena Norberg-Hodge has been prominent in bringing the fate of the Ladakhis to the attention of the West [272]. The relevance for our discussion here is that, in Ladakh, we have an isolated society that has survived for many centuries in an inhospitable region of the world by living holistically. Food resources are limited, so they had to be managed thriftily if the people were to survive. Because their diet too includes little fruit, few vegetables and much salted meat and butter we would expect a very unhealthy population. In fact, although people do get ill and infant mortality is high, the people are very fit, often into their 80s, as they would need to be to do the work their survival requires. This indicates that dietary requirements may not be as prescriptive as medical authorities in the West lead us to think and may be a function of genetic background. Life expectancy overall in the population is lower because the lifestyle is very demanding, but the work involves strenuous exercise in the fresh, unpolluted open air and the food is all organically grown without the use of toxic pesticides. The weather is harsh and the working day long, at least in the summer-time; but they set about their tasks with a relaxed attitude. They work cooperatively instead of competitively, which is the norm in most aspects of contemporary Western living.

There are two morals to be learnt from this story. The first is that the western capitalist dream is not appropriate for everyone everywhere on Earth, and we in the West are already seeing our dream turning into a nightmare. The second is that social cohesion through cooperative living is more successful in producing long-living contented people than individualistic self-serving competition, even when the material comforts are few and the living sparse. Somewhere there must be a middle way to bring some of the advantages of western technology to underdeveloped people to ease the burden of their lives without exploiting them and totally annihilating what was a happy, efficient and self-sustaining society.

Summary

We need a degree of power in order to be effective in implementing change. When this power is abused by a religious group, government or dictatorship, it is not in the best interests of a society or the individuals in it. Such a society cannot flourish. Sociologists and psychologists agree that the pursuit of happiness is the primary objective of human life, embracing as it does pleasure and fulfilment as well as the avoidance of pain. This individual fulfilment is most satisfyingly achieved by participation in the fabric of a society.

A cohesive society cannot be established or sustained without the active involvement of the individuals within it. Rights and responsibilities must be considered in equal measure. Each society has its own characteristic framework and conditions of one society may not be appropriate to another. Individuals, in moving from one social system to another, must be prepared to embrace those of the new society in which they live and cannot expect whole societies to change their ways to accommodate them.

These ideals of social responsibility need to be introduced to children by their parents from the earliest age and then form part of the educational systems in our schools. The difficulties experienced in western society today are at least partly due to the erosion of the extended family unit and its replacement by materialist attitudes. The advertising industry is persuading people to ever greater materialism through the purchase of goods they don't need and can live more fully without. We are brainwashed into believing 'change' is equivalent to 'progress', but the pace of change all around us has outstripped our physical, mental and spiritual evolution.

The fact that some of the observations in this chapter on the state of society were made many centuries ago but are still at least as relevant today indicates that, while we have progressed materially to a huge extent, the underlying social fabric is no more cohesive, and probably less so because of immigration, than it was for previous generations. We seem to have learnt little of social benefit in the last two hundred years. Unless we can act now as a society and internationally against these damaging influences, the future for humankind in the 22nd century – if any – is bleak.

CHAPTER 6

ECONOMICS

There are two main duties of any government. The first is to maintain the rule of law so that its citizens are kept safe. The second of these objectives is economic management – to keep a country solvent so that it is in a position to trade, which provides food and other commodities for the population. For a government, it means balancing the income from taxes with expenditure on social projects. For a company it means maximizing profit for shareholders and company directors for minimum expenditure on staff and raw materials. And of course, we all have to do this on a smaller scale ourselves in managing household expenses against income, but we are concerned here primarily with national and corporate economies.

There is nothing intrinsically immoral about money or the concept of profit. There must be a system whereby individuals can be rewarded for their expenditure of time and effort and for their skill acquired through training and experience. In less sophisticated indigenous societies, the reward may come through reciprocity from work performed by others with specific expertise in another area. In the western system, individuals use the money they obtain to provide essential commodities like food, clothes and shelter for themselves and their families, and use the remainder, if any, for luxury items, such as material goods or recreation. Trading locally is a more socially cohesive and holistic approach than utilizing large impersonal corporations: there is great satisfaction in using local tradesmen or suppliers, building mutual rapport and trust, and knowing that we are directly helping to consolidate our own community. Local tradesmen, conscious of maintaining their reputations, are also more likely to do quality work as opposed to workmen who come from further afield who may not be concerned with reputations locally.

The problem in Western society today is that money has become equated with power. This has always been so to some extent, but the influence of that power is now much greater than in any previous period of human development. Individuals in government and industry now have the power to greatly influence the lives of ordinary citizens both nationally and internationally, and even to influence the environment with their decisions. The kings of biblical times, the pharaohs of Egypt, the senators of Rome all had both wealth and power. The aristocracy of European nations and America distributed a certain amount of largesse to artistic creativity, and we must be very thankful to them that they did, but the greatest part of their wealth was spent on self-indulgence. But these were relatively few rich and powerful individuals, and the mass of the people were at roughly the same level of wealth. It is true that there were also in these societies those who, through sickness or sloth, were forced to be beggars and who did not possess any significant level of wealth, some living in abject poverty.

Today, for far too many people, money has become their master rather than their servant, their objective in life – a life where success and happiness is measured by quantity of possessions instead of by depth of contentment. Those who already have plenty want still more. Materialist consumption has become a symbol of wealth and power. The range of luxuries on offer has widened enormously since the Industrial and Electronic Revolutions and people are judging themselves and others by the size of their houses, the number of cars they possess, how many exotic foreign holidays they take each year, and in other material ways. Money and the quantity of material acquisitions that it buys have become an intoxicating objective of life. A person who lives lightly and simply, who does not drive a top-of-the-range car, who lives in a modest house, who holds no company directorships or high government office is not one whom most people today would describe as successful. Yet a life lived in contentment devoted to helping others wherever possible, making maximum use of whatever talents we are blessed with is surely the most successful that we can live.

The advertising industry exists to create envy and discontent. Like religion, it sets up models of perfection that we can never attain. Trying to assuage this by engaging in 'retail therapy' provides little satisfaction that is short-lived. This consumerism is the source of the alienation we see in the 21st century – a loss of meaning to life because material acquisition is the overriding goal of existence: alienation from one another then arises through our awareness of the gross wealth differentials. This philosophy of consumerism has also brought about a more fundamental alienation from the universe that gave us birth and that nourishes us, physically and spiritually. Because of the drive to continually increase earnings, we now live to work instead of working to live.

This materialist philosophy of life with its social divisiveness and alienation has been encouraged by the increasing creation of lotteries and other gambling systems or establishments. People are enticed by the prospect of obscene levels of prize-money, amounting to many millions of pounds or dollars, for doing nothing other than scratching or writing a few numbers on a piece of card. Giving people rewards – and not small but very great rewards – for doing nothing whatsoever by way of constructive work is morally reprehensible and totally unholistic. In lotteries, a society is rewarded to only a limited extent from the small portion of income distributed to worthy causes. Of course, the organizers of any such gambling institutions will be sure to see that they are handsomely rewarded. Lotteries would not be so ethically deplorable if the rewards offered were more limited and more equitably distributed with a much bigger percentage of the funds directed to worthwhile projects that governments claim not to be able to afford. The rewards are now so high, and the prospects of getting something for nothing so enticing, that many become addicted to gambling as they do to drugs [137, 138]. When the dream of wealth is unfulfilled, people are locked into a spiral of decline through increasing debt, while the moneylenders get richer. The existence of lotteries is just another facet of contemporary society's preoccupation with the acquisition of money as the predominant goal of existence. It encourages the deplorable philosophy of getting instant gratification for virtually no effort. Yet we tell our children that they must work hard in order to succeed in life: little wonder that so many feel the effort is simply not worthwhile. Furthermore, the spectre of addiction is always looming.

All-embracing materialism surrounds us now in western society wherever we look. And it is the job of advertisers to make us want even more, to feel inferior or that we are depriving our children if we and they do not have the latest fashions in everything. Those working in a corporate situation are not socially accepted if they do not conform to these materialistic ideals. This is not to suggest that we could or would want to go back to living in straw huts, with no sanitation or many of the other amenities that modern western living provides. It is possible to live comfortably, taking advantage of modern facilities, and to still live in an ecologically sound way, making time for the spiritual component of life. This is what living holistically means, and the *tao* of holism is the application of this philosophy in our own lives as an example that we hope others may want to follow.

Motivating holistic living is the self interest that we are benefiting our own health and wealth, as well as that of others, directly. We are beyond the point of letting ourselves be cajoled into buying products or pursuing a lifestyle because we think or are told that as individuals we're worth it. We are all aware now of the extent to which we are all interdependent in what has been called the 'global village'. Commerce in the world embraces international trade to an extent that vastly exceeds that of earlier generations. There has always been international trade, sometimes over great distances, even in the time of the Egyptians and ancient Greeks, but the quantity and range of goods traded was minuscule compared to what we have today. With the invention of computers and of the Internet, and tele-transmission by satellite, we now have almost instantaneous communication too. With such global communication and holidays facilitated by air travel we are much more aware than ever before of the conditions under which people in other parts of the world live – at least, we have some sort of picture to replace the almost total ignorance of most of the population in previous generations. It is this global concern that will have to replace the individualistic and corporate lust for wealth and power if we, our children and grandchildren are to survive.

One encouraging sign of a global shift in priorities is the strength of protest, often consumer-led, that is beginning to emerge against large dehumanized American-owned companies and international organizations like the International Monetary Fund, the World Bank, and the World Trade Organization. It is unfortunate that these legitimate and essentially peaceful protests have on several occasions been hijacked by anarchists and turned into violent demonstrations. Those whose protest is sincere will usually demonstrate peacefully if passionately.

There are many points of concern about this global ultra-materialist consumer philosophy.

(i) The gap between rich and poor, between well educated and poorly educated, both as individuals and as nations is increasing rapidly. This divide between rich and poor has steadily grown and, with it, discontent and resentment and a lowering of living standards of the majority of citizens in Britain and America. This undercurrent of discontent may well be a key factor in the increasing amount of vandalism and contempt shown to authority by the young. We look to our leaders to lead us into a more equitable society.

(ii) The profit levels disclosed by many oil companies, supermarkets and other major companies are totally unjustifiable as returns on investment. It is not uncommon to hear of percentage profit increases in double figures, while ordinary citizens are subjected to swingeing double taxation (substantial taxes levied on goods and services paid for with already taxed income!).

(iii) With 'targets for growth' of at least a few percent expected of any company or national economy, environmental resources are dwindling at an alarming rate, and many major environmental problems are being created as a result (see Chapter 7). The Earth cannot afford annual growth of any materialist organization beyond 1%. There are not enough Earth resources for the developing nations of the East to come close to achieving the materialist profile of the West and especially the profligate lifestyle of America.

(iv) With attention focused unswervingly on the materialist capitalist objective of ever-increasing profits, the welfare of the individuals in the workforce producing those profits has become of only secondary concern. In Britain, for the past two decades, management has received annual salary increments well into double figures while the workforce has had to be content with increases of 2-4%, thus widening the already large salary differences. In many countries, by employing part-time workers, who receive no paid holiday leave, no paid sick leave, no pension entitlement, and no health insurance, companies are saving considerable sums on staff costs to the detriment of the workforce.

(v) Time for leisure, for reflection and introspection, and for family contact is being absorbed on the treadmill of the work necessarily undertaken for physical survival. Marriage breakdown means that each potential partner must fulfill all activities necessary to live themselves, instead of being able to share responsibilities. The increasing need for more extended commuting from home to office, which is brought about by centralization of commercial output, extends the working day, again leaves less leisure time, and uses up more environmental resources.

(vi) The disadvantages of commuting apply to goods as well as people. In the present economic philosophy, it increases Gross National Productivity (GNP) or Gross Domestic Productivity (GDP)[1] if certain goods manufactured by one company or country (let's call it A) are shipped from A to B, and identical goods manufactured by company or country B are shipped to A. This is a crazy situation with profligate squandering of resources. If we wage war, we can employ more people making tanks, guns and munitions to kill people, so GDP increases. In peacetime, if there are more road accidents, more people are employed sorting out the tragedies, fixing insurance claims, repairing damaged vehicles, and so forth, and GDP again increases. There must surely be a more sensible way of assessing a nation's economic prosperity than the present system of GDP/GNP, which is morally offensive, socially meaningless and environmentally damaging.

Unless these trends can be halted and reversed, the quality of life for the majority will decrease, the quality of life for the rich and powerful in government and industry will continue to spiral upwards out of control. In the longer term, humankind on Earth will simply perish, whether by famine or other increasingly devastating natural disasters, or by war over access to resources like oil, or even those as fundamental as living space, clean air or water. The disturbing signs of this trend are already with us.

Small is beautiful – global economic factors

One far-sighted economist who wrote, three decades ago, of the impending catastrophes that would result from unrestrained growth in size and power of companies was the German-born economist Ernst F. ('Fritz') Schumacher – and we have gone some way along the predicted path since then. He wrote an inspirational book on the subject of corporate materialist expansion in the 1970s [324]. Some of the comments in the opening chapter of Schumacher's book set the tone of the book: 'Modern man does not experience himself as a part of nature but as an outside force destined to dominate and conquer it' [*ibid.*, p.11]. The undesirable consequences of the growth of mega-corporations formed the core of E.F. Schumacher's book. In Schumacher we read: 'Wisdom demands a new orientation of science and technology towards the organic, the gentle, the non-violent, the elegant and beautiful' [*ibid.*, p.27]. This wisdom was presented in a book that was written more than thirty years ago. What Schumacher was advocating was the philosophy of holism.

Acknowledging that the aim of human existence is the pursuit of pleasure, Schumacher finds that all too often 'to work is to make a sacrifice of one's leisure and comfort, and wages are a kind of compensation for the sacrifice' [*ibid.*, p.45]. This is why we should strive to find work that is agreeable and satisfying and which does more than provide an income for ourselves and our dependants and, more especially, does not involve commuting that damages the environment or the wellbeing of others.

Later in the book, Schumacher deplores the greed, envy and lust for power and profit of 'anonymous companies or megalomaniac governments which pretend to themselves that the whole universe is their legitimate quarry' [*ibid.*, p.29], and he quotes Mahatma Gandhi: 'Earth provides enough to satisfy every man's need, but not for every man's greed'. Today, even more than thirty years ago, it is true that: 'If an activity has been branded as uneconomic, its right to existence is not merely questioned but energetically denied' [*ibid.*, p.34]. At that time, the ultra-materialist periods of the governments of Ronald Reagan and Margaret Thatcher were still to come!

As detailed in the book by Alfred Chandler of the Harvard Business School [48], American mega-corporations have grown to such an extent in size and power over the past century that they now exert an enormous influence over, if not effectively control, the actions of democratically elected governments – and thereby not only in their own country but in the whole industrialized world [see also 19; 27, pp.59, 60]. Governments simply cannot afford to ignore the economic consequences of denying these international corporations almost anything they demand, from favourable tax incentives for development (and thence the opportunity to make still more profit) to ignoring as far as they can the environmental pollution and desecration that many of these enterprises cause. Chandler studied the economic histories of the 200 largest manufacturing companies in America, Britain and Germany, regarded as the three leading industrial nations, during the century from around 1870/1880. The end result of their corporate policies was that they were able to establish near monopolies in the fields of the products that they manufactured, which reduced consumer choice and control. The substance of Chandler's study is neatly distilled for non-economists by Paul Ormerod in his book, *The Death of Economics* [276]. One result of, or perhaps motivation for, this economic expansion in America has been a six-fold increase in their population over the last century, while that in Britain, which also has absorbed a large number of immigrants, has only doubled.

The profit criterion of viability of any organizational enterprise, referred to by Schumacher, was applied ruthlessly in Britain and the United States in the 1980s during the years of the administrations of Margaret Thatcher (elected 1979) and Ronald Reagan (elected 1981). Although thankfully their governments are now consigned to history, their economic philosophy is still applied in the west even to those institutions that we apparently mistakenly regarded as providing some of our fundamental rights in a civilized society, like the right to health care and an education. Not only manufacturing industries but even service organizations like these are now expected to show annual productivity and profitability. Even after nearly three decades of application of these monetarist principles, it is still far from clear whether an increased *number* of examination passes in schools or patients treated in hospitals really is a measure of anything, despite the passion for 'league tables' of performance. American students have had to 'work their way through college' for decades. Now, few university students in Britain will leave after graduation with debts of less than £15,000. If they take longer courses, the debt is likely to be significantly higher.

Cutbacks in education to improve 'profitability' have meant that pupils are now groomed with specific types of examination question to secure passes rather than be *educated* in the fundamental principles of their subjects. Examination results take no account of the quality of the pupil intake – whether from an affluent or socially deprived area, whether pupils are native English speakers or foreign immigrants – or the quality of the teaching environment. The current educational fad for coursework to contribute to examination marks gives misleading results of educational success, since there is no way of knowing the extent of the contribution to the submitted material by teachers, tutors, parents or the Internet. For this reason, there are plans to phase it out in certain subject areas in Britain (2006).

Similarly, in the area of health provision, there is no account taken of the social environment in which, say, a hospital operates, and therefore of the basic state of health or age distribution of the local population in assessing its profitability and effectiveness. Cutbacks in healthcare spending and the obsession with privatization of services has meant an increase in the number of people leaving hospitals with infections that they did not have when they entered. In Britain, privatization of hospital cleaning staff and the abolition of matrons whose responsibility it was to manage hospital systems directly involving patient care meant that the profit motive again became paramount. MRSA (multiply-resistant *Staphylococcus aureus*) became endemic in this period and this has now been joined by another potentially fatal infective organism, *Clostridium difficile.* In the USA, the Chicago Tribune of 23 July 2002 reported an analysis of the results of an extensive survey of patient databases from 5800 hospitals and 75 state and federal agencies that showed that there had been over 103,000 deaths from hospital infections. The period covered by the survey is not given, but the figure probably relates to one year. Even medical authorities recognize that the American healthcare system is the third leading cause of death in the USA after heart disease and cancer [163] and some surveys say it is *the* leading cause of death [275].

Consideration of the health of western nations leads us on to consider the economics of the food supply. The large agricultural multinational companies and supermarkets demanding cheaper food from suppliers are increasingly, directly or indirectly, driving small producers and farmers out of business [155, 156]. Farmers are also encouraged to develop a monoculture of just one or two crop varieties, making them more dependent on their sales outlet and decreasing the planet's diversity and consumer choice [396; 273, p.37]. Crop rotation of grain or vegetables, fallow pasture and regenerative crops like clover in successive years has worked for countless generations and provided the nation with wholesome food. But synthetic farming maximizing profit for the huge agrochemical industry has now become the operating principle in agriculture.

The so-called Mad Cow Disease (bovine spongiform encephalopathy, BSE) was first detected in 1986 in Britain, again during the Thatcher era. Cows, which are herbivores, were fed remains of other dead animals in ground-up feed because it was more profitable. Cattle put on weight more quickly on high-protein diets. As a result of BSE, more than 175,000 cows were killed in Britain alone [92]. The human form of the disease, Creutzfeld-Jacob Disease, which originates by eating meat contaminated with BSE, had already claimed over a hundred lives before the end of 2001 [66], despite assurances from government ministers that meat was quite safe to eat. Intensive farming of poultry led to the prevalence of salmonella in chickens and eggs, so that it became unsafe to eat raw eggs in any product. The Minister in the British Department of Health who made this fact known to the public, Edwina Currie, was promptly persuaded to resign in 1988. The incidence of listeria contamination in meat and dairy produce nearly doubled in the UK in the period 1987-89 [223], because many kinds of monitoring agencies found their budgets slashed in the Thatcher era. The incidence of food poisoning cases reported in England and Wales rose more than five-fold in the last two decades of the 20th century [Facts and Figures of the Public Health Laboratory Service, www.phls.co.uk/facts]. Now officials will be keen to reassure you that this is only because of more effective reporting and recording. Perhaps. But it is more likely that standards have fallen in hospital care and animal husbandry in the rush for profits while the number of officials to monitor standards has been reduced 'to save costs'. If animal husbandry had been conducted according to holistic principles, and human welfare had been a primary concern, these situations would not have arisen.

The Trade Union movement in Britain, which the Labour government under Jim Callaghan had failed to control in the 1970s, was curbed by the Tory government under Margaret Thatcher. Unfortunately, it wasn't just curbed but made totally impotent. As a result, employers had free rein to deal with employees as they wished. There was much wrong with the trade unions as organizations to secure acceptable conditions for their workers, as pointed out by Michael Shanks [330], but their total emasculation was far from an ideal solution. It disempowered employees to too great an extent and gave to the managers of business and industry some of the ruthless power of the autocratic mill and mine owners of the 18th century. Increased spending by the New Labour government in Britain, which abandoned many of the socialist principles of Old Labour in an attempt to court private investment, has so far failed to rectify the immense damage done in so many areas by nearly two decades of application of the philosophy of the government of Margaret Thatcher, whose premiership was one of the most morally offensive and socially damaging that Britain endured in the 20th century.

As a result of the policies of the Thatcher government, the notion of a fair day's pay for a fair day's work, the idea that those who work hard will be financially rewarded while those who do not will not be rewarded, that society will support the disadvantaged, and that basic institutions like the health and education services would be available for the benefit of all have disappeared. In 1979, when the Thatcher government came to power, just under 8% of the potential work-force had no family member in employment: by 1995, that figure had risen to 20% [133]. People now work far longer hours in Britain than anywhere else in Europe and get fewer statutory holidays. There is also a higher percentage of working mothers, either from choice or necessity, and of teenage pregnancies in Britain than in other European countries. This removes the element of personal choice for mothers to be intimately involved in the rearing of their children. As Sue Palmer says, 'if our culture is to have a future we need to create family-friendly economies, not economy-friendly families' [285, p.155].

It was during the Thatcher years that the policy of privatization of health services, education, the rail network, and utility supplies was implemented in Britain, with the result that everyone, and especially the lower-paid and pensioners, have found their living standards dropping ever further. The idea of privatized utilities in Britain was 'sold' to the public (not that we had any choice!) on the grounds that people would be able to 'shop around' and choose the cheapest suppliers; but energy costs have risen continually in all of them, far above the overall rate of inflation. As a result, more than 30,000 people are believed to have died from cold-related illnesses during the winter of 2005/6 because they could not afford to heat their poorly insulated homes adequately. There is no reason why homes built in Britain over the past fifty years should not have had adequate wall, loft and roof insulation, double glazing, and sensible plumbing that saved energy. But the British government has failed to introduce legislation that would ensure that homes were properly built for the pathetic reason that it would be 'an unwarranted intervention in the market' [263, p.65]. It also would have put even more homes beyond the financial reach of still more people in Britain. Such legislation there is, is not enforced and builders know this.

The social damage inflicted by Thatcher's monetarist policies in Britain and, by analogy, those of Reagan in America, is the essential message of Will Hutton's book, *The State We're In* [182]. Thatcher's policy of individualist materialism was the very antithesis of holistic living. But then, she did not believe that such a thing as a society even existed, just collections of self-serving individuals [365]. Because people were described in this way, this was soon how they behaved – and were encouraged to behave. The economic statistics of British companies during the 1980s speak for themselves: investment rose by 2% per annum, profits rose by 6% p.a. but dividends jumped by 12% per year

[182, p.8]. Hutton, echoing the philosophy of Schumacher's book, points out the immorality of the privatized companies now providing essential power and water supplies in Britain 'The directors of these natural monopolies ... granted themselves share options at keen prices which have made them millionaires many times over' [182, p.7]. They still charge customers inflated increases on prices annually to ensure lucrative returns for directors and shareholders on the pretext of using the money for environmental conservation.

Referring to the situation in Britain only a decade ago Hutton states: 'Only around 40 percent of the workforce enjoy tenured full-time employment or secure self-employment; ...another 30 percent are insecurely self-employed, involuntarily part-time or casual workers; while the bottom 30 percent, the marginalized, are idle or working for poverty wages' [182, p.14; see also 191, p.72]. Only the top 30-40% of the population have seen their standard of living rise; for the remaining 60-70%, it has fallen because annual pay increments have not kept pace with inflation. The situation is likely to have deteriorated further since then. Some employers do not even pay staff the legally binding minimum wage saying that employees are too poorly educated to be worth it – and this, despite government statistics recording increased examination performance each year for nearly twenty years!

The fact that material inequalities, both among inhabitants of the nations of the West and between nations of the West and those of Asia and Africa, are steadily increasing is a measure of the failure of the capitalist Free Market philosophy to bring prosperity to all. That the meetings of international finance controllers of the G7 nations are continually disrupted by anarchic demonstrations is due in no small measure to the incompetence, incomprehension and mistaken forecasts of the World Bank, documented by George and Sabelli, amongst others [120]. The British economist Richard Layard believes that this capitalist competitiveness is detrimental to societal cohesion: 'Our fundamental problem today is a lack of common feeling between people – the notion that life is essentially a competitive struggle. With such a philosophy the losers become alienated and a threat to the rest of us, and even the winners can't relax in peace' [216], a view shared by John Gray [134].

Former Soviet president, Mikhail Gorbachev, is convinced 'that the road to finding solutions is blocked by the lack of political will' [129, p.3]. Electioneering to satisfy the personal ambitions of politicians and the materialist consumerism of the electorate wanting to keep capitalist politicians in power are the real reasons why we are not making progress towards a more stable and equitable world. Extremist world-views, like Marxism and fascism in the early 20th century and now Islamic fundamentalism, attract adherents because people living in poverty are desperate to change the outlook of unheeding politicians.

Although the above comments relate to Britain, many of the principles of this monetarist economic philosophy were inherited from America and its resulting problems are spreading throughout the industrialized world. This capitalist ethic, as we shall see shortly from the lesson of Ladakh, is imposing its will at every opportunity on the developing countries. The mega-corporations are infecting the world with their ruthlessly intrusive capitalist philosophy, from Ladakh to the remote islands of Scotland [245; 267; 272]. Because governments of both 'left' and 'right' political viewpoints have now been brainwashed into following these capitalist theories – unproven and in some cases demonstrably incorrect or inappropriate for the existing social structure – everywhere 'there is a loss of faith in state intervention, regulation and ownership as instruments of economic improvement' [182, p.15]. If private companies can run railways, hospitals, schools and utility supplies at a profit, while paying out large sums of money to directors and shareholders, there is no reason why state-run organizations should be unable to do so, so that these huge profits are ploughed back into the state economy. The answer may lie in the reluctance of governments to invest in long-term projects, for each government in a democracy knows it has only four or five years at most to get results if ministers or congressmen want to hold onto their jobs and salaries. With so many systems in Britain and America in a state of decay after years of neglect without adequate financial support, it is now considered beyond the capital reserves of their governments to undertake such an overhaul of public services.

Homes in the northern climates of Canada and Scandinavia have been built with adequate insulation to protect against cold weather for most of the past century. It has taken too long for builders in Britain to begin to apply the same quality standards to house building. Successive governments have ignored this energy issue, which was brought into sharp focus by the oil crisis of 1973/74 and by the publication of the Limits to Growth studies of the early 1970s [250]. Herman Daly showed that the energy conservation situation hadn't improved but indeed had deteriorated further two decades later [71-73]. The looming world energy crisis was only one of the problems these authors anticipated. They also discussed 'the complex of problems troubling all nations: poverty in the midst of plenty, degradation of the environment, insecurity of employment, economic disruption' and other issues [quoted in 276, p.35]. The conservation of energy issue has been neglected by governments for far too long, and still there is no consensus for action.

Economists see the world as a machine, operating through the processes of supply and demand, oiled by the availability of individual and national capital, with an end-product of material growth. Ecologists and environmentalists on the other hand view the world as a living organism where, once essential material needs have been satisfied, the goal of all worldly systems should be spiritual growth. The aim of capitalist economics is to see that the few – the directors and shareholders that it regards as the entrepreneurs that keep its life-blood flowing – are richly rewarded for their faith in the system. They achieve their goal through constant advertising that persuades individuals to buy things that they do not need and often do not even want, because their social status depends on it. The aim of ecologists, in contrast, is a holistic one – that all of society should be involved in and rewarded by whatever developments occur in the course of human societal evolution. It is societal groups that should drive human development, not the few self-serving individuals controlling mega-corporations and the stock market exchanges. The American sociologist, James Coleman, presented an interesting study in which he tried to blend these two opposing approaches to our existence within society [59].

Other economists are trying to find ways of preserving the capitalist system with ecologically sustainable growth. Paul Hawken founded several natural food companies in the 1960s and 1970s and has suggested ways of implementing a 'restorative economy' [158-160]. So all is not lost, by any means. But more people must embrace the new economic ethic if we are to survive. As the former American Democratic presidential contender Al Gore said, the environment must become 'the central organizing principle of civilization' [130, p.269]. This is a message Gore has been disseminating world-wide in his book and lecture-film *An Inconvenient Truth* [131] since his narrow and controversial defeat by President George W. Bush.

The lesson of Ladakh

Ecologist Helena Norberg-Hodge first arrived in Ladakh in 1975 as part of a German documentary team. She heads an organization called The International Society for Ecology and Culture and has worked on their behalf in Ladakh for more than a quarter of a century [272]. Ladakh is a trans-Himalayan region of Kashmir whose administration is divided between Hindu India and Muslim Pakistan, though the predominant faith of the Ladakhis is Tibetan Mahayana Buddhism. Until the late 20th century, the way of life of the Ladakhis was much as it had been for centuries – arduous but fulfilling and, because of their isolation, lived totally in harmony with Nature. After the intrusion of western capitalism, ostensibly to make life 'easier' for the Ladakhis but in reality exploiting another opportunity for profiteering, the coherence of their society disintegrated, as the following tale will tell. Indigenous societies have lived for millennia in harmony with their environment, and think and feel and work within this spiritual unity.

It is thought that the people of Ladakh are of the same ethnic origins as the Hopi and some other North American Indian tribes who are now believed to have originated from the region of central Europe around Lake Baikal and the Gobi desert. The Indians were able to cross into the Americas over a land bridge across what is now the Bering Strait. It is a two-thousand-year-old culture that has eked out a sparse living in the highlands by thrifty use of the available resources of land and water. Hardy crops (barley, peas, turnips, potatoes) and animals (sheep, goats, Asian cattle, yak) are all that could survive under such harsh conditions. But these animals could supply meat, milk, butter and cheese for food, hides for clothing, pulling power for transport, and dung for fuel. Until quite recently, all the essential needs of these simple people for food, clothes and housing were supplied locally. Houses were built and resources were shared cooperatively. People adopted attitudes towards one another that would avoid causing offense or friction, as conflict was alien to their philosophy of life. Because of the close-knit nature of their society there was a profound sense of security. The wellbeing of the individual was truly for the good of the whole society. People could control their own lives. They represented a model of holistic living, not only in their ecologically sound use of resources but in their spirit of happy collaboration in all that they did. It was a harsh lifestyle but they quested for nothing: theirs was the epitome of Buddhist living.

The Ladakhis share with the indigenous tribes of the Americas the same lack of ego – not to be confused with lack of self-respect – and they have many tribal healing and other practices in common. An animal is never slaughtered without prayers for forgiveness. In the philosophy of *ahimsa*, all living creatures are sacred. Whenever a new house is built, there are always prayers to the *sadak*, the spirits of the earth, for protection of the builders and future occupants. There are many words in Ladakhi for 'time', which is not thought of in the western minute-to-minute representation of the clock face but as segments of natural cycles. Dawn is called *nyitse*, 'sun on the mountain tops', while the onomatopoeic *chipe-chirrit*, 'bird-song' is the early morn, before the sun has risen, when the birds sing [272, p.35]. For all their hard work, the Ladakhis still led a more leisurely life than westerners, and the spiritual and rational aspects of life were considered complementary. Although we in the West spend fewer hours a day actually working at our jobs now than we did a century ago, commuting time and stress have increased, and sociologists tell us that westerners spend more hours overall per day working than people who live in agrarian rural economies.

But, the intrusion of Western capitalism has destroyed this ethnic way of life, which has now become competitive instead of cooperative. They now have to grow crops to provide cash income instead of food for subsistence and barter. Unfortunately, the western way of life seems very attractive when viewed from the outside by people living in primitive conditions. They only become aware of the many problems as they start to arise when such a lifestyle is forced upon them through indoctrination by their economic masters. For example, instead of ploughing the land by hand, with a plough made from local wood and pulled by a yak fed on local pasture, the Ladakhis now use tractors. The tractor does the ploughing job much more quickly of course and this seems an attractive alternative. But tractors have to be maintained mechanically with spare parts that cost money. They also run on fuel that must be purchased. Instead of separating the grain from the chaff by traditional winnowing, there is now a diesel-powered mill in Leh, the capital, that is much faster than the traditional water-powered mills. But the inhabitants must carry their wheat or barley many miles to the mill and the processed grain back again – which takes much time, effort and money.

These changes have simply produced a different kind of labour that is less satisfying for the labourers but makes the mill-owners richer. It is better, they are told, to build houses of more durable cement, which of course must be purchased from western suppliers, than to use the traditional cheap local mud. The new developments do not have the unqualified benefits they are promoted as having: the new, faster technologies do not in the end save either time or effort, and they are both socially and environmentally damaging. As Norberg-Hodge says: 'The old beliefs were based on a description of reality that emphasized the unity or dependent origination of all life, whereas the new scientific perspective emphasizes its separateness'; there is 'a shift from ethical values that encourage an empathetic and compassionate relationship with all that lives toward a value-free "objectivity" that has no ethical foundation' [272, p.109]. When indigenous peoples in Ladakh and elsewhere are forced to rely on western companies for supply of materials and the technology that goes with them, their self-respect is diminished and a captive market of increasingly profitable returns is created [324, p.163]. The workforce must then work to the rules set by the mega-corporations rather than in their own way and at their own pace.

Perhaps the most insidious and destructive aspect of western capitalization is that the children of Ladakh and elsewhere in developing countries are now taught that the old ways were bad, uneconomic, backward, primitive, but that the new western technologies will be bountiful and bring prosperity to all in the richness of their lifestyle. After a couple of centuries of the development and expansion of capitalist industrialization in the west, we know this to be false. Expanding education in breadth and depth for as many people as possible is a laudable goal, but not if children are inculcated with the ideas of a competitive capitalist technological society instead of an essentially harmonious agrarian ecological society.

Education of primitive peoples and providing them with some mechanical labour-saving devices and western medicines can bring great rewards in improving their quality and length of life. But extensive and fundamental social change must be done gradually with the wellbeing of the community as the prime objective and we must not allow societies to be exploited to fill the pockets of capitalist entrepreneurs. People used to a life using only the simplest tools and ecologically based agricultural techniques need proper instruction if they are to safely adopt more sophisticated methods of farming. Some 70% of the pesticides used in India are either banned or their use is severely restricted in the West. Farmers are encouraged to use such compounds even when there are no pests! Of course, many pesticides kill not only insects but their natural predators, so where there were no pests before, now there are, and the need to use toxic chemicals supplied by the agrochemical industry can be clearly demonstrated. Field workers in India and Africa, often unable to read any warnings on containers written in English, spray these poisons about on their own and our food without adequate protective clothing. But then, human welfare is an irrelevance in the pursuit of profit. Instead of encouraging these people to grow the food that they themselves can consume, it is more profitable to encourage them to develop a monoculture for export. Thereby, they become dependent on trade in the international market, which is totally under the control of western governments and their financial institutions.

The lessons from Ladakh are clear. The West cannot impose its own lifestyle and values on the peoples of other cultures without their having to pay a very heavy price indeed, both socially and environmentally, for whatever benefits they may receive. The driving force is always an attempt to impose uniformity of economic philosophy that, far from creating unity, is exploiting the environment to unsustainable levels and creating divisions both within and between the nations of the world. As an administrator from the United Nations Development Program comments: 'The world has become more economically polarized ... if present trends continue, economic disparities between industrial and developing countries will move from inequitable to inhuman' [quoted in 191, p.139]. The imposition of western religion has had a similar deleterious effect, creating enemies within formerly harmonious communities and nations like Sudan and Ladakh on the basis of adherence to man-made religious dogma. Most 'ethnic cleansing' is in fact perpetrated in the name of religion and attempts to impose religious uniformity. A holistic lifestyle for individuals or for nations cannot be achieved by such measures. Individuals and nations must be free to follow their own spiritual ideals and lifestyles appropriate to their culture and tradition.

Is the Free Market really 'free'?

The whole structure of capitalist economics in the West is based on the notion of a so-called Free Market. The Scottish political economist Adam Smith (1723-1790) published his *Enquiry into the Nature and Causes of the Wealth of Nations* in 1776 just as the Industrial Revolution was getting underway, and in the same year as the American Declaration of Independence was signed. It was he who established the philosophy of capitalism that developed and progressed henceforth to the present day with its laissez-faire attitude – that 'market forces' will always determine the success or failure of any enterprise, without any form of governmental interference, and that wealth will always 'trickle down' from the rich to the poor. Smith was an idealist and believed that the benefit that accrued to any individual would be a benefit to society as a whole. Unfortunately, Smith took no account of human nature – that the rich and powerful generally want to get richer and more powerful rather than squander their wealth on raising the living standards of the poor and disadvantaged. We met this striving-for-superiority psychology described by Adler and the will-to-power of Nietzsche in Chapter 2, and the philosophy of continual questing for wealth and power described by Thucydides and Hobbes in the previous chapter.

Since the Industrial Revolution, people have increasingly been treated merely as *working* beings rather than as *human* beings. Yet the purpose of the Industrial Revolution, people were told at the time, was that mechanization would give them greater leisure time. In fact, as a result of this particular social revolution, people were forced to do tedious, repetitive and often more dangerous jobs instead of their previously fulfilling, if arduous work. They were frequently employed in making just a part of a useable object instead of the whole object, which was then sold on at a profit to someone else, thereby depriving them of the satisfaction of a completed task or the utility of the article they had helped to make. And most importantly, in every sphere, 'time' became a commodity with capital value, usually coupled with 'labour' as 'man hours'. The stated objective of increased leisure had evaporated with the gleam of increased profits. Many 19th century writers deplored this enforced slavery of the working classes, notably Karl Marx (1818-1883) and Charles Dickens (1812-1870). With the continued exploitation of the workforce since their observations, we now have a state of materialist imperialism to replace the 19th century territorial imperialism.

The economic collapse of America in the late 1920s and of Germany in the 1930s that brought Hitler to power, the oil crises of the 1970s and 1990s, and the three-day working week in Britain in the 1970s has made economists aware that this laissez-faire capitalist philosophy does not always work. The Free Market philosophy has been adopted as incontestable economic truth even though there is ample empirical evidence that it has very many failings and is not the panacea for economic ills that it is made out to be. As David Jenkins, former Bishop of Durham, says: 'A world picture is being constructed which is treated as if it were a revelation about our world . . .We are invited or compelled to accept this world picture as gospel truth . . .It is 'gospel' because it is good news – it shows ways to flourish' [191, p.xi]. Capitalism has become the new opiate of the masses: 'The theories we are sold about the Market, and the promises we are told we may or must invest in it, are much more like stories or myths than convincingly articulated scientific theories' [191, p.2]. Capitalism or Free Market economics is the new *Geist*, the Hegelian view of societal ethos that controls the way society behaves. We are encouraged to have 'faith' in the system – indeed, we are given no choice. In a publication from 1995, the *Financial Times* columnist Sir Samuel Brittan admitted that 'most of the hard core economic doctrines are . . . ways of looking at the world' [34, p.17], just like the various religious scriptures. These economic doctrines are not logical statements (as in mathematics), nor are they theories or laws based on and verified by empirical observations (as in science), but simply working hypotheses that have been shown by many incidents of financial crisis to be inadequate. Their advocacy is a form of economic fundamentalism: 'this simple modern model of the Market is taken as being literally and universally true over all human communities and environments, and timelessly true into the indefinite future' [191, p.118].

The Market is treated as if it were a machine, always behaving in the same way with utterly predictable outcomes. We know just from the economic disasters of the 20th century that the Market behaves much more like the Earth as *Gaia*, as an unpredictable organic interacting whole. Human society and its economic structure are constantly in a state of change, just like the ancient philosophers said of the natural world. But sages like Heraclitus and Siddhartha Gautama also pointed to the interconnectedness of all that is, and in the globalization that has taken place increasingly over the last half century this is certainly true of our economic interdependence. Crises in the Nikkei stock market index in Japan, or failure of the tea crop in Sri Lanka cause reverberations worldwide, and problems in oil supply are even more catastrophic in their global effects. Much of the food for the developed nations is now imported from overseas so no countries are economically self-sufficient, especially when acquisitions and mergers put essential resources of one country under the control of others.

We have long since reached the stage where the economics of the stock market determines the fate of governments and individuals rather than independent election by the people. As pointed out by Ormerod, economists prefer private to public ownership because it gives them a greater measure of control: they turn a blind eye to the fact that they are really engaging in political manipulation [276, p.48]. Factors outside of the stock market, and beyond their control, can have a devastating effect on people's lives and thence rebound on the Market. Economics may embrace timeless principles, but the basic resources on which those economic factors depend are of only limited availability. It is foolhardy to believe that we can continue talking of 3% annual growth or more in economies in a world of finite resources.

So great was the success of *The Wealth of Nations* that it almost totally eclipsed another earlier work by Smith, his *Theory of Moral Sentiments*, published in 1759. In this Smith had emphasized the importance of 'sympathetic feelings' between employers and the workforce in the emerging industries to ensure that what we would now call 'job satisfaction' was maintained. These workers were doing mundane, repetitive jobs often under appalling conditions in increasingly grimy industrial centres. They had formerly been merchants and craftsmen living albeit hard-working lives but of some quality in rural areas. Some employers took note and created schools and built housing for their workforce. Too many however, enchanted by the prospect of vast profits from the labour of others, the basic ethos of capitalism, exploited the workforce and established a precedent that has now become normal practice to this day. It is the financial benefits of directors and shareholders that are always paramount over consideration for the welfare of employees, the principle being that large profits establish confidence and lead to growth. Consideration of quality of product and customer satisfaction also is entertained only insofar as it maximizes profits.

A reorientation of this attitude is fundamental to holistic living. Just how this can be achieved in capitalist economies, where even services like health, education and the provision of basic amenities like water and power have become profit-oriented, is hard to see. Unless a majority of the people can be convinced of the immorality and impracticality of the present system and encouraged to voice their disenchantment at the polls by election of responsible leaders, the situation will continue to deteriorate. We have returned effectively to a feudal system where the money-and-power brokers hand out beneficence in wages instead of land.

Adam Smith brought out new editions of both of his seminal works, with alterations and additions, throughout the remaining decades of his life, so he clearly regarded the moral welfare of society as important an issue as its economic welfare, with the two prospering together. Likewise, the financial benefits to the individual, though pursued out of self-interest, were somehow thought to benefit society as a whole, though this was not the overt intention of those who invested their capital: 'He generally, indeed, neither intends to promote the public interest nor knows how much he is promoting it . . . he intends only his own gain and he is in this, as in many other cases, led by an invisible hand to promote an end which has no part of his intention' [345, IV, ii, 9]. As David Jenkins pointed out, this Invisible Hand is like some universal *Geist* operating a beneficent 'overall picture of the world and of human life, morals and political and economic activity within it' [Jenkins, p.44]. It is like divine Providence working through humankind for the fulfillment of creation [191, p.87], as envisaged in A.N. Whitehead's Process Theology [193, p.63]. It was believed that somehow 'the system', that is, Free Market economy, would operate in *everyone's* best interest. But as we have seen, both within and between nations, this simply doesn't work as the divide between the rich and poor is continually increasing.

The root of the problem is the operating philosophy of those who run the capitalist system of the West, forcing companies to continually expand and engulf small operators, to maximize productivity and profitability without heed of human or environmental cost, and imposing western ideas of democracy and 'growth' and Christian individualism on all societies and countries in the developing world [268]. Living in a self-sustaining, ecologically sound local system of demand and supply is 'not economically profitable'.

There is a huge expenditure of energy involved in the transportation of foodstuffs around the globe in order to implement this Free Market economy. This point was made emphatically by Helena Norberg-Hodge and her collaborators in their book *Bringing the Food Economy Home* [273]. For those food items we do need to import we should see that those who produce the goods at source receive a decent financial return on their labour. An encouraging sign of our global holistic development is illustrated by the number of 'fair trade' outlets that are now arising.

In his *Small Is Beautiful* book [324], Schumacher deplores the extensive use of chemicals in agriculture to boost productivity, the consequences of which on desecration of the environment were so graphically predicted by Rachel Carson [46]. But there are direct material human costs as well as the moral issue of poisoning our wildlife. It is estimated that up to 70% of the antibiotics that are used in the developed western world are fed to farm animals as growth promoters or as a preventative against infection – to which they would not be exposed in the first place if they were managed organically. Helena Norberg-Hodge tells us [273, p.59] that 'roughly half the 25,000 tons of antibiotics produced in the United States are used in the raising of animals for human consumption', and most of this is for growth promotion or disease prevention as opposed to disease treatment. Public health officials are in no doubt that the inefficacy of so many antibiotics now in humans is due to the amounts ingested in human food from animal sources. Whatever platitudes government officials dispense to assure the public that there is no risk to them, significant traces of these chemicals do pass through the food chain. Such an abundant use of antibiotics in the rearing of animals gives the bacteria more opportunity to develop resistance. The prevalence of debilitating if not fatal infections like MRSA (multiply-resistant *Staphylococcus aureus*), *Clostridium difficile* and Campylobacter are a clear indication that in some cases the bugs are winning! Tuberculosis was thought to be almost eradicated in the West as a fatal disease until a new mutated strain called MDRTB (multi-drug resistant TB) emerged in the 1990s. Like MRSA, it does not respond to even our most powerful antibiotics. Furthermore, the use of factory farming methods and long-range transport of food back and forth for packaging and distribution from increasingly large supermarket depots is thought to be responsible for the increasing number of cases of food intolerance and food poisoning from salmonella and listeria [100, quoted in 273, p.62]. We should not forget that the introduction of diseases common in the civilized West were fatal to indigenous natives when western explorers and their armies invaded the Americas. It was disease rather than military conquest that vanquished these people. We now have our own debilitating or lethal diseases to contend with as a result of our capitalist life-style.

Seeds are now being genetically modified so that they will force the use of more pesticides, in order to increase company profits with no concern for human or environmental welfare, which are an irrelevance in today's unholistic psychology within the multinational companies. Monsanto has genetically engineered seeds to produce crops that will be tolerant to even larger doses of its best-selling herbicide, Roundup; and Monsanto are not alone in pursuing this exploitative philosophy. The agrochemical company Aventis produces its own seeds that respond to its own herbicide, Liberty, while Cyanamid does the same for its herbicides Pursuit and Odyssey [273, p.48]. The pesticide chemical glyphosate (the active ingredient in Monsanto's Roundup) is promoted as being 'of low toxicity' to humans; yet it is acutely toxic to all animals and is suspected of causing some forms of cancer and other adverse effects in humans [64; 148]. The organophosphate insecticide malathion is another commonly used pesticide with many adverse effects reported on humans [65]. Chemicals too, like the pathogens, are alien to our human biochemistry and are debilitating the human constitution to an ever increasing extent.

Field trials of genetically modified crops are underway in several European countries, but containment is almost impossible, as seeds or pollen are distributed far afield by wind, birds and insects. In America, genetically engineered foodstuffs are now extremely common because manufacturers have used a combination of senate lobbying, litigation and intimidation to see that they can be incorporated into foodstuffs without any warning to consumers of their inclusion. We all need to be mindful of the fact that this 'improvement' in our food supply is irreversible: once genetically engineered crops are growing, the process cannot be undone.

Perhaps some of these altered foodstuffs will do us no harm though, as I said in Chapter 1, there is evidence that normal bacteria in the human gut are being altered by genetically altered foods [166; 268]. The genetically modified varieties are not superior to natural varieties of crops in nutritional value: they simply allow crops to be grown under climatic conditions for which they have not naturally evolved.

The immorality lies in the fact that we have no choice as to whether or not we consume genetically modified food. In America, manufacturers and TV personalities are prevented by law from making any comment that foods might not be free from contaminants and are safe to eat. Oprah Winfrey was charged under these laws for daring to make a disparaging remark about beef [153, p.129]. The U.S. has no laws however to prevent fertilizer from containing toxic waste or radioactive materials. News of these dangers and other contaminants in our environment is difficult to obtain since the news media are owned or effectively controlled by the corporations who want such public service information stifled [12]. The incidences of poisoning that began to be reported so frequently in the 1970s from the toxic waste that had been dumped at the Love Canal site, Niagara Falls, New York, for half a century alerted the American people to the extent of uncontrolled toxic waste dumping that had occurred nationwide in the interests of fiscal expediency with the full knowledge of local authorities.

The capitalist ethic now has such a stranglehold on world economics that governments – and the common people – feel that There Is No Alternative (TINA) [182, p.66; 191, p.9; 245, p.2]. Government ministers and company executives alike see the Free Market as the only way forward, and look to it with optimism for the bounty that it will surely provide. In the decades following Adam Smith's publications, it was a common belief that the Free Market, once established, had a self-perpetuating momentum through the influence of the divine 'Invisible Hand'. This economic philosophy, once it became established, has flourished to the present day. Capitalism depends for its survival on 'growth', on continually increasing the material demands of consumers by building in obsolescence or increasingly limited lifetimes for as many products as possible [279; 367, Chap.4].

No government has yet shown the courage to grasp the nettle of ecological sustainability or to confront the thorny problem of electability with realistic policies. The prospect of developing nations like China, India or countries of the African continent, with their burgeoning populations, trying to secure for their people the kind of profligate lifestyle of the Americans will clearly exhaust Earth's resources long before the end of the present century. David Jenkins comments on 'the cynical collusion between multinational companies and corrupt governments which slants trade to the detriment of the majority of the inhabitants of too many developing countries' [191, p.98]. Meanwhile, the living standards of ordinary people in developed countries are slow to improve and, in some cases as we have seen, are actually declining. All is not well even in the mega-rich countries of the West: 'the rich barricade themselves in fortified ghettos while the poor exist in increasing numbers of 'no-go' areas' [191, p.99; see also 267]. With the collapse of the communist system in Eastern Europe in 1989-91, the capitalists of the West were jubilant. Here was the vindication of TINA – that theirs was the only economic system that could produce a stable society. Actually, it is just that its downfall is somewhat longer in coming!

But there is hope of a new outlook. For example, economists Herman Daly and Robert Constanza founded, in the 1990s, what has become known as the International Society of Ecological Economics (ISEE) with its Secretariat located outside Washington, D.C., and they launched and edited a journal called *Ecological Economics* with a publication office in Boston, Massachusetts. The European Community has a comparable branch of its own called the European Society for Ecological Economics (ESEE) with its own journal, *European Values*. So there is an influential body of academics that takes this environmental situation seriously.

There is a dichotomy of fundamental beliefs at issue here. Do we stimulate material growth for the few through the Free Market, hope it will 'trickle down' to the masses and thereby produce happiness for all? This is the philosophy of the 'right-wing' politicians and most contemporary economists. Or do we redistribute wealth as evenly as is practicable through taxation (or total state control in the case of communism) to produce a more equitable society believing that a more contented people will work harmoniously together for growth and their common prosperity? This is the policy of the political 'left'. In fact, neither system can be implemented in practice in totality. In the era of the ultra-individualist policies of Reagan and Thatcher, only the power brokers at the top of the social hierarchy prospered while the majority of people were dispossessed and resentful. In the communist system, working for the state destroyed all personal incentive. Despite the glee with which the fall of Soviet communism has been greeted in the west, far from defeating the Marxist philosophy that inspired communism, capitalism has actually embraced the Marxist concept of growth through materialist consumerism. Our 'sacred sites' are now the supermarket and the shopping mall, our 'religion' is consumerism based on the 'scripture' of Free Market economics, and the Earth is the sacrificial offering.

According to the Austrian economist Friedrich A. Hayek (1899-1992), a nation must embrace one philosophy or the other, capitalism or socialism: there is no compromise. However, some economists, like H.J. Witteveen, former Managing Director of the International Monetary Fund, and those of the ISEE mentioned above believe that it is possible, indeed, essential, to start educating people into a more spiritual and humanist philosophy of life [397, 398] even while we live in a capitalist system, and moderate our materialistic behaviour. Hayek was in no doubt which of the two economic philosophies was superior, as capitalism was held to be based on sound, rational, scientific (that is, empirical) principles [161, Foreword]. Many economists strongly disagree and see capitalism as no more than a self-perpetuating custom [182; 191; 276]. The other side of the capitalism versus socialism argument to Hayek is presented by financier and political philosopher George Soros, who believes that Free Market capitalism is to be feared more than communism: 'I consider the threat from the laissez-faire side more potent today than the threat from totalitarian ideologies. We are enjoying a truly global market economy in which goods, services, capital and even people, move around quite freely, but we fail to recognize the need to sustain the values and institutions of an open society' [350]. Soros was born in Budapest in 1930 and had to endure first the German then the Soviet occupation of his native Hungary. He escaped to be educated in the London School of Economics and then settled in America, so he has experienced both extremes of social and economic philosophy at first-hand. While few would want the kind of extremist socialist state exemplified by the USSR or China, the dangers and flaws in an ultra-capitalist economy like that in America are apparent for those who wish to see [191, p.105; 267].

Both capitalist and communist socio-economic systems are gargantuan. For the sake of humankind and planet Earth, we must make politicians find a compromise and one that moves more in the direction of Nature spirituality and away from adherence to dogma, whether political, religious or economic.

Summary

We have seen that, in capitalism, the rich just want to get richer and more powerful, so 'trickle-down' of wealth rarely happens, and we have seen the psychological motivations behind these actions. If we try to implement too much state control, the system becomes unworkable, overburdened with bureaucracy, and personal incentive is destroyed. If we cannot look to either capitalism or communism, we need economists and politicians to work together in harmony for the good of the people who elected them and find a realistic compromise economic system that exemplifies the principles of holism with a more ecologically sound lifestyle.

What we will eventually have to return to for our very survival is a set of numerous small local economies. Gone will be the massive systems for international and transnational transportation of food. Aircraft will, for the most part, be grounded unless a new means of powering them can be found in the next few decades. We will be relying mostly on local or home-grown fresh food, grown using traditional organic farming methods instead of depending on soil being force-fed a continual cocktail of toxic chemicals. Gone will be the farming systems sustained only by the agrochemical industry, which will be moribund if not obsolete. The soil will be nourished again by plant and animal waste, and its natural wildlife, poisoned for decades by fertilizers and pesticides, will be restored.

Our health will improve as a result of eating more nutritious food and people will gradually take responsibility for their own health. Thomas Edison was perhaps being overly optimistic when he said: 'The doctor of the future will give no medicine but will instruct his patients in the care of the human frame, in diet, and in the cause and prevention of disease', but the thought was inspired. Naturopathic medicine will again predominate as the need for allopathic medicine from the pharmaceutical industry diminishes. Gone will be the automatic prescription of toxic drugs for every minor ailment. Herbal remedies and alternative medicine will predominate, spiritual healers will be in demand, the pharmaceutical industry will be scaled down (many of its products are derived from coal and oil!) and everyone will be much healthier anyway from eating food that once more has both flavour and nutrients. The tobacco industry will not be able to survive commercially as increasing numbers of people realize the wisdom of not smoking. All of this will also benefit individuals financially, so a way must be found to stabilize national economies in the new world order.

We shall be content to take short holiday breaks at destinations no further away than those accessible by our electric cars as the petrochemical industry becomes extinct for lack of oil. Gone will be the massive schools of over a thousand pupils, transported by car and bus from miles around. Children will go to school locally on foot or by cycle, and as all pupils will be commuting in the same way, there will be adequate safety for all. With healthier food and more exercise, the epidemic of childhood obesity in Britain and America will be curbed and eventually reversed. Is this a Luddite vision of a return to a lifestyle that embraces the simpler values of former centuries or the Way to a prosperous and happy holistic future for humankind? We shall see.

CHAPTER 7

ENVIRONMENT

In the 1960s we knew about the dangers of pesticides to wildlife and to our own health [46], but the use of pesticides and fungicides has increased considerably since then [155, 156; 273; 342]. We knew that resources of natural fuels like coal, oil and gas were likely to run out in the 21st century, but few governments have made any substantial commitment to research on renewable energy sources. We knew that many natural resources were becoming depleted, but we were very tardy about starting a programme of recycling. We saw that global temperatures were rising from our production of 'greenhouse gases', but governments have done little to reverse the process, despite the efforts of prominent campaigners like Al Gore [130, 131]. We knew that pollution from the industrial world was damaging the environment of countries not involved in spewing out these toxic chemicals into our air and water supplies. It has taken us too long to begin to get to grips with these problems. Still, in some countries, there is not even a national programme to deal with these issues let alone international agreement and cooperation. Still, greed for profit is everywhere the predominant motivator. Let us consider some of the issues.

The natural processes of change

Most natural processes work in cycles that are in a state of dynamic equilibrium with one another, that is, they are stable overall but in a state of constant change. The cycles of growth, decay and re-growth in the plant world are evident to us all during the seasons of the year. The days, the seasons, the years themselves are cyclical, and we measure our concept of time by the periodic motion of sun, moon and stars. Rocks are formed from sea-floor sediments washed down by rivers from the land or from the skeletons of marine organisms deposited on the ocean floor.

These are then compacted by pressure; they are raised up into landforms by crustal movements, and wind and water weather away the structures; the fragments and particles are carried away by wind and water to lie once again as detritus on land in structures such as screes or desert sands, or as sediments on the sea floor. Water evaporates from the sea and rises into the atmosphere. When the water droplets coalesce, they fall as rain or snow, which runs into rivers and then back out to sea again. The atoms themselves are constituents of this cyclical process, and they may often change molecular partners as they move within these natural cycles around the Earth. At the subatomic level, all matter is continually cycling back and forth with energy. Inorganic systems generally move in the direction of increasing disorder or greater entropy. These cyclical processes are superimposed on the continually evolving universe.

The most important of the atoms for living systems is carbon (denoted by C), which is an indispensable component of all life or organic structures. Atoms of hydrogen (H) accompany carbon in all organic structures. Another important atom is nitrogen (N), and we shall explore shortly how N and C move in cycles in Earth processes. The other important atom in living systems is oxygen (O), which animal tissues need for respiration and which is incorporated into the carbohydrates that comprise the starches and cellulose in plants. Oxygen is the most abundant of all the elements, for it makes up some 21% (by volume) of the atmosphere, nearly 90% (by weight) of the water of the oceans, it is a constituent of the carbonate, silica or chitin that make up the hard parts of sea creatures, and it is found abundantly in rocks as silicates. C and N form the molecules that comprise the backbone of the protein and nucleic acid polymer structures of all animal tissue. The atoms of C, H, O, and N (with the help of some sulphur and phosphorus) shuffle around in cycles between different partners in the processes we describe in living organisms as metabolism. With a constant input of energy, these processes are responsible for maintaining plant and animal life. Organic systems grow by using energy to develop into a state of greater order or lower entropy.

Natural processes of change tend to be relatively slow and rhythmical, so that Man and other animals and plants can adapt and live in harmony with these environmental changes. The changes produced by Man have been much more drastic and are increasing exponentially with population. We seem to have become afraid of standing still, lest we be thought of as being out-of-touch or old-fashioned and 'change' has become synonymous with 'progress' [367, 368]. This unconscious feeling of fear is a great cause of stress. Man-made change is all around us today, but as E.M. Forster remarked: 'How can man get into harmony with his surroundings when he is constantly altering them?' Man is suffering from what Edward Goldsmith calls 'cognitive maladjustment to the industrial world that he has created' [127, p.436]. Humankind is not genetically programmed to deal with this mechanized and electronic world that we have created over just the last few centuries, which is so far removed from the world of trees, rocks and streams among which we have evolved over a few million years. We have lost our awareness of spirituality in the pressures of modern living.

The nitrogen cycle in Nature

We hear much in the media at present about the carbon cycle, but very little is said or written about the equally important nitrogen cycle. Nitrogen (N) gas makes up most (about 80%) of our atmosphere, but in this form it is essentially inert, unreactive and insoluble in water. Nevertheless, it is an essential component of the structures of every protein and nucleic acid molecule found in all living tissue. We get our useful N from eating plants and animal tissue, but the ultimate source is atmospheric N. Plants too cannot use the gaseous form of nitrogen but they do possess the means to turn it into a useable form.

Natural soil is pervaded with myriad organisms. Moles and voles make their homes in the earth beneath our feet, churning it up as they do so. On a smaller scale there are earthworms, roundworms, millipedes and beetles. Smaller still we have bacteria, blue-green algae[1], protozoa (that feed on bacteria) and soil fungi[2]. The food for these micro-organisms is decaying vegetation, like leaf litter, and animal waste. The importance of this cycle for human nutrition is that, as part of their participation in the N cycle, the micro-organisms make the minerals in soil available to plants and thence to us in our food. Without the living organisms, no matter how 'rich' the soil may be in humus and minerals, plants will never be able to take up the nutrients into their systems. The rotation of crops and the leaving of fields fallow for grazing in successive years have been practiced successfully by farmers for millennia, without any knowledge of agricultural biochemistry.

In the latter half of the 20th century, the chemical industry discovered a new way of making money. On the pretext of producing more food to feed the world's hungry millions they started producing agricultural chemicals, with a massive promotion aimed at farmers and governments concerning the advantages of their use. By encouraging farmers to use liberal quantities of a cocktail of pesticides and fungicides, the agrochemical industry succeeded in wiping out not only weeds (whose natural decay products would have been a valuable source of nutrients) but also most of the living organisms in the soil. Then, farmers needed even more liberal quantities of fertilizers to make anything grow at all on the soil. For the animals, there were generous helpings of antibiotics, just in case they developed infections, and growth hormones to help them put on weight at an artificially fast rate. Thus, a new agricultural cycle was established. American agrochemical manufacturers are so powerful that they have used their financial resources to see that government passes laws banning dairy product producers from mentioning whether or not these chemicals have been fed to the cattle from which the milk is obtained [153, p.63]. There is considerable evidence that natural farming produces tastier, safer and more nutritious food at a much lower cost [77, p.190; 156, Chap.7; 273, Chaps 3 & 4; 290; 304]. It is true that the quantity of food produced has increased hugely in recent decades. The downside of this is that the food has become increasingly tasteless and lacking in essential nutrients, because they are simply not there in the soil and, even if they were, the plants could not assimilate them.

The utility of trees

As discussed in Chapter 4, trees and shrubs and other vegetation have always had a symbiotic relationship with humankind. They have provided us with food, and with building materials for our homes and for ships to sail the world. Until the development of the pharmaceutical industry in the 20th century, medicine men have relied for countless centuries on grasses and the leaves, nuts, berries, and the fruits and flowers of trees and shrubs to treat our ailments[3]. If we think about our daily food intake, most of it is supplied directly by the vegetable kingdom – our breakfast cereal and the soya milk we use to moisten it (and even the dairy milk, meat and poultry come from animals fed on seeds or grass, hopefully), the toast made from wheat flour, the vegetables for our dinner, together with the nuts, seeds, soya or quorn we use as source of protein if we are vegetarians, the fruit we eat for dessert – all of it from the vegetable kingdom or what the scientists call flora (a term that includes much more than just flowers!). The plant kingdom supplies all of our food, directly or indirectly. Then there is the aesthetic joy of simply walking in the shade of a woodland on a hot summer's day, especially if it is carpeted with wild flowers. In Chapters 1 and 2 we met child psychologists who are stressing the need for children to spend more time playing in green areas. If they were to spend more time as children in parks and woodland learning about trees, flowers, insects and birds they are more likely to respect and treasure these features of our world as adults. The practical issues of maintaining our green and wooded areas have become paramount, as we shall see.

The natural carbon cycle and the effects of human ecological disruption

Apart from its providing us with food, herbal medication and spiritual uplift, a further practical aspect of our need for green vegetation is its production of oxygen. This became a focus of attention globally in the 20th century. Green leaves take the carbon dioxide that we produce by breathing (and by the burning of fossil fuels!) and, through the process of photosynthesis, produce oxygen. This is the reverse of our process of respiration, which involves breathing in oxygen to metabolize sugars for energy and expelling CO_2. The trees retain the carbon to make cellulose and starch for their structures and food. The whole oxygen–carbon dioxide system is in equilibrium — or would be within wide limits but for human interference with this robust natural cycle.

When tropical forest is cleared for human agriculture, as has happened in recent decades in South America and Asia, the environmental consequences are enormous. When trees are felled we are deprived of a vital CO_2 sink and source of oxygen; the forest is often burnt to clear it 'efficiently', thus adding still further to the CO_2 build-up; a habitat for wildlife and indigenous peoples is destroyed with untold threat to numerous other species of plants and animals; our moral obligation to preservation of the natural ecosystem is flouted and we reduce biodiversity [396]; possible sources of food and drugs are extinguished; and often soil cohesion is destroyed, leading to soil erosion, lower fertility, and clogging up of river systems, destroying their wildlife too. All of this is without considering the aesthetic beauty of the forest we have lost. As Thomas Berry says: 'We have silenced too many of those wonderful voices of the universe that spoke to us of the grand mysteries of existence' [27, p.17]. We think of the Earth only as source of raw materials for industry or object of scientific investigation rather than as a world of wonder and mystery that could provide us with unending delight: 'We need to move from our human-centered to an Earth-centered norm of reality and value' [27, p.56]. To think that we are enhancing human quality of life by destroying the environment in which we live is an illusion. The current destruction of tropical forest for agriculture by developing nations must be halted, whatever resources the developed nations have to expend to get the job done because the impact is global and permanent.

The temperature and atmospheric carbon dioxide levels of the Earth have varied over the course of its history. We do not know whether the present increase in CO_2 levels is due in part to a natural global cycle, but there is no question about Man's role in this and it has become essential to limit our own contribution and minimize the largely undesirable effects for humankind of a rise in global temperature. There are few people now who do not know of 'global warming'. There is still some scientific controversy on this issue, but the overwhelming majority of research scientists in the field are convinced that there is a real threat from increased CO_2 levels that are being brought about largely by human activity.

In the face of such evidence, one has to consider that intelligent people opposed to this idea, like Ian D. Clark of the University of Ottawa, Canada, and Bob Carter from James Cook University in Australia [125(a)], or American eco-journalist Ronald Bailey [15, 16], have a hidden agenda in trying to minimize the dangers of global warming and adhering to this ecological nihilism – perhaps they are associated with or sponsored by some major polluters, they are enjoying their hedonistic pleasures and do not want to see them curtailed, or they are religious fundamentalists who believe that, whatever we do to planet Earth, God will sort it out for us. The American economist Julian Simon (1932-98) was another who was in denial about exponential population growth, resource depletion, and global pollution; he advocated exploiting Earth's resources relentlessly, ignoring the fact that they were not infinite [340]. Such selfish stupidity and irresponsibility seems almost incomprehensible in the face of the potential ecological catastrophes that confront us. The situation is not helped by reports from some journalists who demonstrate their ignorance of the mass of scientific assessments by taking a controversial stance against these warnings. Thus journalist Peter Hitchens reports: 'The greenhouse effect probably doesn't exist. There is as yet no evidence for it' [125(b)], which is simply not true.

The American journalist David Helvarg showed in his writing just how strong was the opposition to any curtailment of capitalist growth through environmental exploitation [164, 165]. This opposition reaches to the highest level of government. Publication of a report on sustainability commissioned by Democratic President Jimmy Carter in 1979 was suppressed by his Republican successor, Ronald Reagan, who took office in 1981. The American coal and oil industries actively block attempts to develop alternative technologies [119; 153, p.27]. As pointed out by Thomas Berry, 'corporations now own or control the natural resources of the entire planet directly or indirectly' [27, p.121]. With profit their only motive, it is not in their interests to have any regard for the environment. In his excellent book on the profit-obsessed materialism of big business, George Monbiot uses the Faust legend as a metaphor for their attitude [263]. Indeed, he devotes a whole chapter of more than twenty pages documenting in detail the lies perpetrated by companies such as Exxon concerning denial of global warming. *Moral issues and human welfare are an irrelevance – profit is the only objective!*

In an attempt to redress the balance of our interference with CO_2 levels, there are measures underway to find ways of capturing and storing CO_2 back into deep rock strata, such as the basalts that underlie much of India, or oil-bearing strata beneath the North Sea. It has been proposed that CO_2 could be stored in ocean floor sediments and strata to replace the oil and gas taken out. There is also the Castor project in the North Sea aimed at capturing CO_2 in a solvent, then separating and storing it for commercial use. But such measures are at an early stage, they are frequently costly and they do not obviate the need to halt the destruction of forests everywhere.

The idea of 'carbon offsetting' is wrong-headed: this entails planting a few trees or doing something else to reduce CO_2 levels if we damage the environment by flying away to holiday in other parts of the world or otherwise increase greenhouse gases. This makes the issue of saving our planet into another trading scheme, like most other aspects of our lives in the materialist society in which we live. The solution is not to take those exotic holidays in the first place and to reduce CO_2 levels *without* doing anything to increase them.

What is indisputable is that humankind is contributing massively to these rising CO_2 levels, whether or not there is an underlying natural trend, whether or not they presage imminent environmental disaster, and whether or not some people on Earth might benefit from the higher temperatures. At the time of the Renaissance the concentration of CO_2 was about 250 parts per million (ppm) of air. That may not sound much, but it had been sufficient to produce and sustain the ordered distribution of vegetation and people over the globe through all historical time. In the 18th century, after the effects of the Industrial Revolution had become evident, it rose to 280 ppm. The latest current figures give an average level of 380 ppm – an increase of some 50% as a result of human activity in just three hundred years. The concentration of CO_2 is now higher than it has been at any time for the last 650,000 years. We know this from the analysis of ice cores from some of the few thick ice sheets we have remaining in the world in Greenland and Antarctica. The increasing industrialization of developing countries like India and China will add to the atmospheric CO_2 production as well as the factor of expanding population everywhere.

The CO_2 in the atmosphere acts like a blanket and traps reflected sunlight to produce a 'greenhouse effect' – a warming of the atmosphere right around the planet. Carbon dioxide from the burning of fossil fuels is not the only greenhouse gas. Water vapour is a natural contributor to the effect, but we need the water breathed out by trees and evaporated from the oceans to produce clouds and perpetuate the water cycle. Another greenhouse gas is methane, produced naturally from decaying vegetation in wetlands but, increasingly, produced by human activity from landfills, in the growing of rice, and as excretion from ruminants: more people means more cattle reared for food. Nitrous oxide, N_2O, is another greenhouse gas, and this too comes from wetlands but more is generated by human activity from sewage treatment and animal manure management and from car exhaust. We must not forget the influence of these other, sometimes more potent, greenhouse gases: reducing CO_2 emissions alone will not slow global warming. America produces 25% of the world's greenhouse gases, and the large SUVs they love to drive produce only about 12 miles of driving to each gallon of 'gas'. The use of rail and canals for carrying non-perishable goods, provision of more public transport, and cycling or walking for us as individuals where possible, will considerably reduce pollution from cars. Several European countries have integrated transport systems of buses scheduled to meet trains and bicycles carried free of charge. Although such a system was promised for Britain nearly a decade ago, it has not appeared. Britain has enough difficulty getting even one privatized fragmented transport system working efficiently, without trying to coordinate one with another. Hybrid gas-electric cars are already available and hydrogen fuel cells are being developed as power source; their widespread use could cut toxic car emissions by up to one-half where car use is high or growing [38, p.127].

This increase in global temperatures is almost certainly responsible for producing the instability and extremes of weather, melting of glaciers and ice caps, and the threat of extinction of more species of flora and fauna, to list just some of the effects. Forecasts of ecological catastrophe if we do not change our ways are not exaggerations [246].

Most glaciers move at a rate of some few centimeters each year. Greenland's biggest glacier is now moving and melting at a rate of 12 km per year! If the whole of the Greenland ice melts it is expected to raise the sea level world-wide by about 7 metres. The 30,000 year old glacier at Glacier National Park in Montana is now melting so rapidly that scientists believe it cannot survive for more than another 30 years. Despite refutation by naturalist and former environmentalist David Bellamy [124], the World Glacier Monitoring Service in Zurich says: 'The latest studies show unequivocally that most of the world's glaciers are retreating' [263, p.24].

With melting of the polar ice caps, the continued existence of polar bears in the Arctic and penguins in the Antarctic is most immediately under threat. If the whole of the Antarctic ice melts it has been calculated that sea levels will rise by 50 metres. The rise in sea level from melting ice has already produced flooding of some Pacific islands such as to make them potentially uninhabitable: the Gilbert and Ellis Islands have signed an agreement with New Zealand for this country to accept all of the island's inhabitants over a period of five years. Melting of the polar ice-caps will also have a devastating but unpredictable effect on ocean currents and world climate. Melting of permafrost in Alaska has already destroyed the foundations of many homes. If the permafrost of the Canadian and Russian tundra were all to melt, it would release an enormous amount of methane, which is an even more 'efficient' green-house gas than CO_2.

The increased air temperatures that are melting snow and ice in polar and mountain regions has a knock-on effect on food supplies. In the Indian sub-continent, the melting of Himalayan ice-fields leads to more precipitation falling as rain that runs off the land causing soil erosion in the rainy season but provides less melted snow to charge the rivers during the dry season depriving farmers of vital water for crop irrigation. A report from the World Bank in 2001 indicated that wells now have to be drilled to a depth of 1000 metres and more in China to reach ever depleted reserves of water at greater cost. As a result, the production of cereals that form staple foods in the diet has been falling for the past several years in China.

The sea is also an important sink of CO_2. Sea creatures use the dissolved carbonate to make their shells, but the sea holds less CO_2 as the temperature rises. So global warming produces higher atmospheric CO_2 levels from this source too, and the higher CO_2 produces more global warming – another natural cycle that we do not want to encourage. Because of rising sea temperatures, coral reefs are dying off. The algae that have lived symbiotically with the coral for many thousands of years cannot survive in the warmer water. Even the coral of the Great Barrier Reef off northern Australia, the most spectacular coral colony on Earth, is under threat. Here we have another formerly stable ecosystem that we are already destroying. Coral reefs may seem of little significance in the materialist philosophy, but all of these ecosystems are important to us practically, quite apart from the aesthetic pleasure they provide, because everything is connected. The butterfly flapping its wings in the English South Downs producing a rainstorm in the Andes is only a slightly exaggerated metaphor for this interconnectedness.

As societies become more affluent their consumption of food generally, and of animal protein in particular, increases. With the health benefits of eating fish instead of meat, aquaculture now provides a substantial source of animal protein in our diets. Water has become a valuable resource on sea and land. Scientists are warning of a complete collapse of the fishing industry world-wide by the middle of the 21st century if present fishing quotas are not curtailed. The collapse of the cod fishing industry off the coast of the Maritime Provinces of Canada from over-fishing resulted in the loss of 40,000 jobs in Newfoundland [38, p.9]. The Peruvian anchovy fishing industry collapsed in 1972 due to over-fishing. This represented some 20% of the world fish catch and it supplied protein meal for the raising of livestock and poultry. This led to increasing use of grain as feedstock and was one of the factors that led to the use of recycled animal carcasses for animal feed in Britain that in turn generated BSE ('Mad Cow Disease').

The human contribution to greenhouse gases can only increase as world population increases, though eating more food derived directly from plants in a vegetarian diet and eating less food derived from cattle would help. Cattle require some 7 kilograms of grain as feed for each kilogram of beef protein they produce, so grain reserves are conserved by a more vegetarian diet. Also, as the economies of developing countries become more affluent, farmers tend to plant higher-value more labour-intensive crops, both for home consumption and export, rather than grain, thus depleting world grain stocks further.

In India, the average consumption of grain is less than 200 kg per year. As the land production is no more than this, most of the grain is eaten directly in flour products. India has also become the world's leading producer of dairy produce. The cow is considered to be a sacred animal, so cattle are bred for dairy products rather than beef.

In Italy, average grain consumption per person is about 400 kg per year, much of it eaten as pasta dishes. With liberal use of olive oil in their dishes, the Italians have one of the healthiest diets in Europe, despite limited expenditure on health-care provision. Each American on the other hand eats, on average, about 800 kg of grain a year, most of it indirectly in the form of animal protein [38, p.41]. As stated above, this is a much less ecologically friendly system because it takes about 7 kilograms of grain to produce each kilogram of animal protein.

Saudi Arabia's wheat harvest has been halved because of aquifer depletion, and wind erosion of depleted soils has similarly reduced the grain crop of Kazakhstan, which experienced the results of over-exploitation of the land, similar to the American Dust-Bowl of the 1930s [38, p.9]. These are not just local problems that those of us living elsewhere can shrug off as irrelevant to us. They are warning signs that if we continue on this course, further and much more widespread and more devastating examples of such tragedies are imminent for us all. Many authors are warning of these events to come but we in the general population must pressure politicians to see that they take measures to halt this destruction [129; 214]. Unrestrained capitalist economics is destroying the very world we live in, the resources of which we need to provide for our survival.

Water shortage is even more critical than the oil situation. We have lived for millions of years without coal and oil, but we can survive only for days without water. We use some 3-4 litres of water a day as actual drinks, but it takes about 2000 litres of water to provide a day's supply of food. As about 70% of the world's water supply is used to produce food, a water shortage will soon become a food shortage. It takes 1000 kilograms of water to produce 1 kilogram of grain. This interrelation between Earth's water cycle, soil erosion and grain production is an example of holistic principles in action.

We need to develop water efficiency in industrial processes, like those in coal-fired power stations that use huge amounts of water. We need efficiency in home use and dispense with machines like dishwashers that also take much water, and use washing-up water to irrigate garden plants. Gardening may soon cease to be as popular a hobby as hose-pipe bans become more frequent, but the use of water-butts to collect rain-water provides a practical, if more labour-intensive, solution. Concrete patios require no water, but paving over garden lawns will deprive garden birds of a valuable source of food. Plants rely on birds and insects for pollination, so fewer birds and insects means poorer crops for food. Furthermore, a concrete patio does not absorb water like soil, so in heavy rainstorms, which we experience more frequently now, there is more risk of local flooding.

Major rivers are drying up, and when they flow through more than one country there is potential for dispute over water rights. The Nile has its source in Ethiopia and flows through Sudan before reaching Egypt, which depends on the river for its existence, as it rarely rains there. The Colorado River was once a major river in the southwestern United States but is now largely used up before it ever reaches the sea. The same is true of the once great Yellow River in northern China which frequently runs dry before it reaches the sea. The Indus in Pakistan flowing into the Arabian Sea and the Ganges in India, which flows into the Bay of Bengal in Bangladesh, both of which are used extensively for irrigation, similarly have little water in them by the time they reach the sea, especially in the dry season because of melting of Himalayan glaciers, as described above.

The Dust Bowl of the 1930s in the Great Plains of the United States, that caused the migration of more than 2.5 million people, was caused by over-development of the land for agriculture. There was a similar situation in the USSR in the 1950s as a result of the Virgin Lands Project to turn grassland into land suitable for growing grain. The same situation is now developing in Africa and China from over-ploughing and over-grazing. In the 1950s there were just five major dust storms in China. In each decade the number increased steadily until, in the 1990s, the number had risen to 23. Natural vegetation reduces or prevents wind and water erosion of soils. About 20% of Earth's surface is forest and another 20% is grassland; 10% is under cultivation but the remaining 50% is essentially uninhabitable mountain, desert or ice. So the arable land and water that we have must be used wisely.

While some regions are experiencing a greater number of tornados and flash floods, in others the process of desertification is marching on relentlessly. This is a significant problem in China, which has 20% of the world's population and the world's fastest growing economy. The Gobi desert is now only about 150 miles from Beijing, and several smaller deserts elsewhere in the country have started to merge. China has a desperate need for agricultural land. In the 1950s, just over 150,000 hectares were lost to desert each year, but by the 1990s the loss had more than doubled to 360,000 hectares annually.

Earth provides a wonderful natural example of holism at work, because our planet functions as an organic interactive whole, as pointed out by James Lovelock in his Gaia books [229, 230]. But Earth has not yet had long enough to stabilize its atmospheric systems to this dramatic and relatively sudden change in conditions – a mere few hundred years out of the nearly five billion years that life has existed on Earth. Locations all over the world are describing weather that is unseasonably hot or cold; excessive rainfall is producing local floods; and storm force winds are much more frequent. As a result, violent weather systems are generated as Earth tries to re-establish its equilibrium. The hurricane Katrina that struck New Orleans in the summer of 2005 generated winds in excess of 150 mph, flooded 80% of the city, killed more than 1000 people and left many thousands more homeless. It was part of a pattern of increasing frequency and intensity of hurricanes in the Caribbean area. It is the rising sea temperatures that are thought to be most responsible for these dramatic atmospheric effects because they affect the flow of air currents over the oceans.

Flash floods produced by severe thunderstorms are becoming more frequent, like those in Boscastle, Cornwall, UK, in the summer of 2004; in central Europe in 1998 and 1999, and again in 2003 and 2005; in southern Alberta, Canada, in June 2005, where three weeks of flooding left 72 people dead; in Mumbai, India in July 2005 that killed over 700 people; and hurricane Allison in June 2001 that killed 30 people in the Houston area of Texas. These few examples illustrate the fact that incidents of such flooding from excessive rainfall in unexpected places or at unusual times of the year are becoming more frequent and more damaging world-wide.

At the other extreme, the drought in the Amazon basin in 2005 killed thousands of fish that were left stranded by dried-up streams. More insidiously, many of the trees are dying still from lack of water in the extreme heat. This is in addition to the trees lost by deliberate deforestation. The Amazon Basin contains 58% of Earth's tropical rainforest (23% is in Asia and the remaining19% is in Africa); these trees produce water vapour to feed the rivers and oceans as well as removing CO_2. To lose much of this is a suicidal disaster!

The death of so many trees means less CO_2 can be absorbed, which again produces rising global temperatures that lead to unstable climatic conditions and more trees in a state of stress. This is a vicious spiral that we must interrupt. In Australia, there is a corresponding increase in the intensity and frequency of forest fires. Forest fires have always been a natural occurrence in Australia and, given time, the forest does regenerate; but again it is this increase in intensity that is worrying. The fires are now much more difficult to control, and there is insufficient time for the forest to recover before there is another outbreak. It would be difficult to halt this process even if the increase in CO_2 levels could be stabilized immediately. The massive forest fires in Indonesia and the Amazon Basin in 1998/99 arose because of excessive logging for profit and deliberate burning of scrubland that burnt out of control. It is not only trees that die in the heat: the hot summer of 2003 in Europe killed some 35,000 people. It is quite obvious that when we harm Gaia, we harm ourselves. We are an integral part of the natural world, not separate from it.

There seem to be few individual entrepreneurs or governments willing to develop renewable energy industries based on wind, solar, wave or geothermal power. The installation of solar heating in homes or industrial buildings in Britain or the USA, let alone in developing countries, is so far very low-key and sporadic. The only alternative energy source not producing carbon dioxide that seems to be seriously considered by the British or American governments is nuclear power. In his book *Small Is Beautiful* [324], E.F. Schumacher also criticizes the development of nuclear power, which once again in the 21st century is being presented as an attractive prospect for energy generation to reduce carbon dioxide emissions. In the West, far too much attention has been focused on nuclear power for energy generation. It can only be made to look commercially viable by ignoring the massive cost overall in developing and decommissioning of power plants, and by ignoring the problem of the disposal of nuclear waste, for which no viable solution is yet available. The effects of the Chernobyl disaster of 1986 will continue well into the next century. Since the partial core meltdown at the Three Mile Island nuclear plant near Harrisburg, Pennsylvania on 28 March 1979, no new construction work has been started on nuclear reactors in the U.S.A. As terrorists become ever more sophisticated and heedless of their own existence or the safety of innocent people caught up in their atrocities, nuclear power plants with their radioactive material transported back and forth over land and sea pose an inviting target. With the money already invested in nuclear energy, this seems to provide a quicker solution to the energy crisis, but the hazards and total costs of nuclear power make it unattractive as an energy source when there are plentiful renewable sources from sun, waves and wind available with technology already developed to use them.

If each home were equipped with solar panels and a small windmill, this could well provide enough power for their household needs and possibly surplus that could be sold back to the power generating companies. Solar panels on large industrial buildings would be equally effective. In northern latitudes they are most efficient on south-facing roofs. Studies of the potential ecological benefits of wind power in the United States have shown that there is enough harnessable wind power to meet even the U.S. electricity needs several times over [5; 38, p.127], but the American coal and oil industries have lobbied extensively to prevent their development. Globally, wind-generating capacity increased eight-fold from under 5000 MW in 1995 to nearly 40,000 MW in the following decade. Europe has played a leading role, with Germany producing a capacity of some 15,000 MW while Denmark now generates 20% of its electricity needs from wind power [38, p.130]. Using low-power light bulbs or fluorescent strip lighting instead of traditional incandescent bulbs may seem a trivial measure, but such lights have five or six times the life of incandescent bulbs and use only about one-fifth of the power – a double saving on household bills and a huge decrease in national energy expenditure. Much electricity is also wasted by leaving equipment in 'standby' mode; unfortunately, however, far too many articles are made to be used in just that way. Fridges and freezers are available now with vacuum-insulated panels so that they use only about one-eighth of the electricity of normal equipment; but trying to find one to purchase is not easy. Manufacturers have a responsibility to see that all electrical equipment is as energy efficient as possible and widely available. There is clearly an urgent need to develop and implement renewable energy technologies nationally, and for individuals to take whatever steps they can to save energy, both to benefit household finances and to save the planet.

The population explosion and its consequences

There are diverse effects resulting from the process of CO_2 increase, as discussed above, and several contributing causes but the most fundamental is the increase in world population. World population stood at a mere 80 million at the time of the ancient Greek philosophers. As a result of wars and epidemics of plague exerting their own population control, this figure had risen only to 700 million by the start of the 18th century Industrial Revolution. We passed the first billion mark in the early 19th century – but we had taken about two million years to reach that level. With improvements in nutrition, sanitation, and treatment of diseases, population had increased to 1600 million in 1900, and it took an interval of only a century to reach a world population figure of two billion (2000 million). The third billion was added in the period 1930 to 1960 – just 30 years – and the fourth billion took only fifteen years. Now, at the start of the 21st century, world population has already exploded to a potentially disastrous 6600 million. We are adding to this figure at a rate of a quarter of a million people each day, 90 million people each year. There are many issues demanding our attention in attempting to achieve a stable and holistic world, but the most urgent factor in correcting *all* environmental imbalances is to restrict and then reverse the trend in world population increase, whatever any political or religious philosophy might stipulate to the contrary [127, p.311; 316, p.238]. It is time that reason took over and we abandoned these highly damaging man-made dogmas.

The population of the world will eventually decrease, perhaps through implementation of some form of control, like Indira Gandhi's sterilization of 11 million men in India (though that was voluntary, eventually it may become compulsory); or by imposition of punitive tax penalties for large families; or from wars like that in Iraq fought over reserves of oil or water or land; or else through starvation as we see periodically in Africa because Earth's limited resources are running out. Perhaps global infections will take their toll. The 1918 pandemic of influenza killed between 60 and 100 million people, and the death toll from the current epidemics of AIDS is still unknown. Disasters have been averted so far from avian 'flu, or the Ibola virus, but viruses have a great capacity to mutate. One way or another, Earth's population will not increase indefinitely and, however morally objectionable, some measures will eventually become necessary unless there are natural events to reduce population, if those now starving or living in abject poverty are to secure any kind of quality human existence.

The desire of eastern developing nations to emulate the material prosperity of the West is understandable; it seems morally reprehensible to say to developing nations that they cannot achieve material prosperity because of our excesses in the West. But the level of profligacy shown by the United States in the past century cannot ever be achieved for the huge populations of India or China, as there are simply not natural resources enough available, quite apart from the pollution factor. The material demands of western nations will be curbed in this current century, whoever sits in No. 10 Downing Street or the White House. Each American uses 25 barrels of oil each year in the home or through air and road travel; each inhabitant of Britain uses 11 barrels per year, while each Chinese person currently uses just 2 barrels per year. Since the first oil-well was drilled in Titusville, Pennsylvania, in 1859 we have become utterly dependent on this natural resource in little more than a century. But, while sixteen new oil fields were discovered in 2000, in 2001 there were only nine new fields, and in 2005, only one. The world's oil supply is running out and is likely to be exhausted in the next fifty years, but oil has become one of the most essential of all raw materials in our materialist 21st century life-style.

It may seem unthinkable to us now, but joy-riding around in motor cars, or flying away several times a year to foreign holidays may become a thing of the past in the not too distant future. We need to start adopting a mind-set now of using car and air travel only when absolutely essential. As stressed by E.F. Schumacher in his book *Small is Beautiful* [324], by Paul and Anne Ehrlich in their assessment of human ecology [90], and by

Helena Norberg-Hodge and colleagues in their book on food distribution [273], we need to start thinking now of moving back into essentially self-contained localized communities, and to do more of our communicating electronically. The trend to centralization of education into schools of over 1000 pupils, so that most pupils have to be driven or bussed in, rather than being able to walk or cycle; and of centralizing food supplies into hyper-markets necessitating a huge transport system to convey food back and forth across the country if not across the globe; and of manufacture of many commodities by mega-corporations that have been encouraged to swallow up small local producers will have to be reversed as oil resources run out. We must remember just how many commodities in today's world are derived from oil, apart from fuel for our homes, cars and airplanes – bitumen for the roads, many common drugs like aspirin, cosmetics, all common plastics and synthetic fibers, paints and dyes, glues, and many, many more products.

Global warming is not the only environmental imbalance that we have produced by our materialistic excesses. Acid rain, produced from burning fossil fuels, and contamination of our environment from toxic chemicals like dioxins and PCBs, now found even in animals at the Poles and in human breast milk in western society, are by-products of trying to feed and provide material comforts for this burgeoning world population and dispose of our rubbish. Our carefree use of chlorofluorocarbons (CFCs) for decades in aerosols produced a thinning of the ozone layer around the Earth and particularly in the atmosphere above the Polar Regions. CFCs were first produced in the 1930s and found widespread application in many household products before it was realized, some fifty years after they were introduced, that they were destroying the ozone layer. One kilogram of CFCs can destroy about 70,000 kilograms of atmospheric ozone by a series of chain reactions brought about by sunlight. Each CFC molecule is also about ten thousand times more effective in its 'green-house' effect than a molecule of CO_2. We do not hear so much about ozone depletion today because we have lessened the extent of ongoing man-made destruction, but it is still occurring and will continue at least to a small extent for many decades yet. The reason we need our ozone layer is to protect us from the ultraviolet radiation from the sun. In small doses, u-v light is beneficial and helps produce vitamin D from its absorption by the skin; we need vitamin D to help us use the calcium in our diet to maintain the strength of our bones. But too much ultraviolet radiation damages skin cells and makes them very susceptible to development into skin cancers. So we need to hold on to the ozone we have left!

The holistic world-view embraces all of the subjects discussed in this book. They are holons within the holarchy of human social interaction. As Mikhail Gorbachev says: 'In all its history humanity has never faced tasks as complex as those we are encountering today' [129, p.55]. We cannot address just one of these holons without affecting others. Gorbachev suggests that the United Nations could convene a 'Council of Elders' of scientists, philosophers, economists and theologians with other Nobel Prizewinners to advise politicians on how best to proceed for the benefit of all. Countries, or corporations within them, should be subject to prosecution and sanctions from the international community if they flagrantly disregard ecological holism.

The list of man-made environmental destruction goes on and on, and affects every inhabitant of every nation world-wide, as the above few examples illustrate, even if some specific event seems geographically remote and irrelevant to everyone else. Any government that does not take drastic and immediate measures to curb CO_2 emissions and develop renewable energy sources is foolhardy and irresponsible, however unpopular its measures may be with the electorate. But power is a great aphrodisiac, and it remains to be seen whether our world leaders have the necessary resolve. Certainly a holistic outlook, or even a rational one not blinkered by political platitudes or corrupted by greed, tells us that we must be concerned at the destruction of woodland and forests world-wide, if only for our own self-preservation. The abandoning of our spiritual roots (Chapter 4), the emphasis on individualism (Chapter 5), and the worship of capitalist consumerism (Chapter 6) have combined to produce the lamentable desecration of planet Earth described in this chapter. *This inconsequential materialism cannot be sustained and any such attempt will simply result in the destruction of the planet and the extinction of our species.*

Diversity is not only the biosphere's right to existence but its preservation is our heritage and responsibility as the highest order of the living species. As we saw in the previous chapter, it is not only natural but also economic and cultural diversity which is being destroyed. From a purely selfish point of view for the human species, the natural world provides us with all of our food, water and air, and has given us a renewable source of herbal medications to treat ourselves when we are ill. We destroy or continue to pollute this living source at our peril! Planet Earth is not an inert lump of rock hurtling through space, like our moon. It is, as James Lovelock suggested [229, 230] a living planet with finely tuned interactive systems that are essentially in a state of dynamic equilibrium or homeostasis until they become unbalanced by the activity of Man. Although we have made a start at restoring the balance, we must increase our efforts to heal the Earth and work towards a much greater global spiritual cohesion alongside the economic globalization we have today. The urgency of the situation has been pointed out by many researchers and writers (see 38; 153; 214].

An anonymous sage said: we have not inherited the Earth from our parents; we have borrowed it from our children. The mystical abbess Hildegard of Bingen (1098-1179) commented, with such foresight, a thousand years ago:

> Everything that is in the heavens,
> on the earth, and under the earth is
> penetrated with connectedness, penetrated with relatedness.
> Hildegard of Bingen (1098-1179)

Let us hope that a sufficient number of those whom we appoint to positions of power heed the wisdom of the mystics.

Summary

The above description gives an account of the practical implications of our not living holistically with the rest of the natural world and what we need to do about environmental degradation. If we provide our environment with conditions that it can adapt to, it will survive. This requires material sacrifices and a good deal of forward planning for all of us in taking care of the environment in all of our everyday actions. We must impress on our governments to take this issue more seriously and develop sustainable methods of energy supply. The basic technology for solar, wind, wave and geothermal energy production already exists but these have not been developed for public use. Of course, this means that present energy suppliers, their directors and shareholders, will suffer commercially and governments are intimidated against doing anything that interferes with the capitalist free market. However, those energy suppliers with foresight are already developing these alternative and renewable energy sources. Until a safe method has been established to dispose of nuclear waste, nuclear power is not an option that any responsible government should consider, whatever its attraction in reducing carbon dioxide emissions.

Fundamental in all of this is encouraging people to live in harmony with others as a global community rather than living as if they were the only person on the planet. Being as sparing as possible with environmental resources like water and energy; not polluting the environment by throwing away rubbish in the countryside or dumping litter around the streets but recycling whatever we can; growing our own vegetables if possible or, if not, shopping locally; buying what are increasingly called 'fair-trade' commodities that give a more adequate return to producers in developing countries; eating more grains, pulses, fruit and vegetables rather than the much more environmentally damaging and possibly less healthy animal products; taking care of our health by exercise and weight control; not smoking, and drinking alcohol only sparingly, if at all; and having respect once more for the spiritual dimension of living: these are all contributions to living holistically that are available to all of us.

Humankind is a part of the natural world and we must therefore work with Nature not against it. Halting exploitation of Nature is a priority not only for ourselves but also for future generations. Living in harmony with others is part of our care of the environment: as the German philosopher Immanuel Kant said: 'Human beings are sentinels on Earth and may not leave their posts until relieved by another beneficent hand' [197, p.154]. Our pollution of the environment, of the very air, water, food and land that we need for survival, is a symptom of our irrationality as well as unconcern for others and for the future for ourselves. We need once again to live in harmony with the trees, rivers and rocks of the natural world to allow us to access the wisdom that lies within them.

CHAPTER 8

EDUCATION

The place to start inculcating ideas of a sustainable ecologically sound holistic existence must surely be in our schools, and as early as possible. The Italian educationalist, Maria Montessori (1870-1952), had the foresight to realize that this process should begin in primary school: 'We shall walk together on this path of life. For all things are part of the universe, and are connected with each other to form one whole unity'. What a wonderful message to convey to little five or six year olds as they begin their exploration of the wonders of the world. Whatever subjects are taught in the curriculum an emphasis on holistic living must feature prominently if humankind is to survive into the next century because so many issues in our society are reaching crisis point.

Sue Palmer's recent book [285], as the title and subtitle indicate, also gives us positive directions in which we can move to create a physically and spiritually healthier environment for our children. Prerequisites for learning are a healthy diet that avoids junk food, adequate exercise and sleep, judicious use of television and computers for entertainment and learning, time spent with parents and other family adults in reading and talking, and adopting a relaxed lifestyle and a holistic moral outlook that avoids any specific religious indoctrination. Many children are being encouraged to live the kind of frantic lifestyle of their parents, with activities of different kinds on every weeknight and at weekends. Lots of activity to use up that bountiful energy in childhood and widen horizons is great, just as long as there are times too for quiet reading and talking [175]. Somehow, space must also be found in our school curricula for activities like sport and music appreciation.

While accountability has become all important in our capitalist-driven western society, the preoccupation with examination results and league tables in the British educational system does no favours to the children. It increases their stress level to be continually working for formal exams and destroys the joy and excitement of learning. It inhibits teachers from straying from formal syllabuses because of time pressures to get all the prescribed work done. Project work is good if it provides the pupils with a chance to use their imagination and explore around a topic. But projects are usually tightly prescribed so in fact there is little or no scope for initiative and, too often, material for projects originates from teachers, tutors, parents or the internet, which might improve exam results but otherwise is of limited use to the pupils. A return to the system of continual informal assessment of pupils by the teachers who know them would benefit teachers and pupils alike. It is also important that teachers, like pupils, should be realistically assessed: very few of those who train as teachers in Britain ever fail as long as they complete the course [363]. If classrooms became, once again, pleasant learning environments instead of combat zones, the welfare of the whole society would benefit.

As for trainee teachers, there is now an educational philosophy that no pupil should 'fail'. But children do need to be given a fair assessment of their abilities so that they can come to terms with their shortcomings in some areas and try to address them. There will be some subjects in which they are strong, and others in which they are weak. To heap praise continually on children even when they are academically weak does them no service at all and simply leads to confusion. If they know that they are not making the required effort, but still 'passing' examinations, there is little incentive to make greater effort. If they are making their best efforts but understanding still eludes them, either the subject teacher has to adopt a new approach or the pupil's efforts must be directed elsewhere. The philosophy that no child must ever fail is misguided. We all fail at times in life and the ultimate purpose of schooling is to prepare children for life. They get enough of the brutal realities of life on their television sets: most of it hopefully will never affect them directly. But knowing there are fields in which they are unlikely to succeed provides an opportunity for them to redirect their efforts into areas where they will be rewarded with success and, consequently, happiness.

Basic literacy and numeracy skills are something that children have to persevere with, as a reasonable level of success in these areas is essential whatever their future employment. As pointed out in Chapter 2, children must learn even at primary level that learning requires effort, that not all rewards are immediate, that resources must be shared, and the needs of others must not be ignored. There is much more counseling of children now than even a generation ago and this provides a nurturing environment in which to emphasize children's strengths but also to address weaknesses and help to build social interactions.

The purpose of education is to give pupils knowledge:

(i) basic knowledge of literacy and numeracy to enable learning to continue – the '3 Rs', reading, writing, and arithmetic calculation;
(ii) general knowledge of life skills, including something about how government, the tax system and stock market work; some science for all arts pupils and some instruction and creative opportunities in the arts for all the budding scientists; some fundamental information about taking care of themselves, like nutrition, hygiene, sex education, parenting and relationships; and
(iii) specific knowledge of the different subject areas to give the child a broad education and help him or her decide which direction their career path or paths should follow [289] and the criteria for choosing a career.

These are the positives. We also must work to eliminate the toxic influences on our children's lives, like fast food, violent video-games, and marketing indoctrination with incessant consumerism.

Education needs to focus on creativity and understanding of the context in which knowledge is to be applied rather than the mere acquisition of facts in order to pass examinations. The human mind requires pegs on which to hang factual knowledge if it is to be memorized in a meaningful way and later used in the very different arena that children will experience when they leave school. Children need to be prepared for the fact that they may not pursue the same career throughout their working lives. In today's market, flexibility in dealing with different employment situations is essential. There is a huge responsibility on parents to do the best that they can for their children and not leave all the responsibility to teachers for their wellbeing in view of the many potentially damaging influences at large on their young lives [114; 232; 240; 285]. Most children are born with insatiable curiosity and a thirst for knowledge. It is this sense of wonder at the world that is just opening up for them that we must nurture.

All children need some workable knowledge of English, if they live in Britain or North America. Without lucid English they do not have the means to express themselves clearly or to comprehend what others are telling them. If we value the beauty and versatility of our language, then we cannot allow its constant slovenly use in the written output of children in the classroom, encouraged no doubt by what is heard and seen continually on television, for this can only lead to decline of language and, with it, of constructive rational thought.[1]

Television 'soaps' present a highly damaging example of human behaviour. A large proportion of the exchanges in these episodes are aggressive with the combatants continually shouting and swearing at one another. We could be spending our time in much more constructive and meaningful activities that would contribute to our happiness and wellbeing [240, p.158]. As Sue Palmer says, 'It would help a lot if television – the window through which we now all see the world – could reflect the advantages of healthy social interaction, rather than concentrating so much attention on dysfunctional relationships and social breakdown.' [285, p.291] As she says, family breakdown has now become so commonplace that it has become a form of entertainment [285, p.148]. If children are led to believe that this is the way differences of opinion are resolved we cannot be surprised if they are aggressive to teachers and fellow pupils. However much the liberals may scream 'censorship', it is difficult to see who else but the government has the power and the duty to influence this if we do not want to encourage a nation of adolescent delinquents. The majority of pupils just want to get on with their learning in peace, but those few who are led astray by these poor role models can disrupt a class or even a whole school.

Computers undoubtedly have a part to play in learning, as long as they supplement and do not replace face-to-face contact and as long as they are used mainly for education. Computer games may provide some educational benefit and entertainment but, if played to excess, these advantages are far outweighed by the loss of human contact and outdoor play activities that children could experience in the time spent on screen. Children need to be encouraged to work off surplus energy and to channel their competitive spirit through sport. They should not be indoctrinated with relentless violence in their video-games or perpetual consumerism in their television. Team sports also encourage the development of the ability to work with other people.

Fortunately, more adults are now becoming aware of the urgency of the need to address this fundamental issue of our children's education for our survival as a coherent society, so there is hope that children will follow the example set by their parents. But time is running out. It matters greatly what kind of knowledge we present to our children, because in a holistic world-view the act of gaining knowledge not only transforms the knower but also that which is known through future interaction with the learner. The more we make children aware of the need for environmental conservation, aware of the plurality of world-views, and the need to live holistically in every aspect of our lives, the more likely it is that these worthwhile aims will be fulfilled when they are adults.

It is generally agreed by teachers and psychologists in the developed world that the education of children is failing badly. Those who recruit school leavers for industry say that their applicants often lack basic skills in literacy and numeracy – so much so that some employers do not consider their workers worth paying even the legal minimum wage. The pressure on parents to work hard enough to provide for a family often leaves less time and energy to play with or read to children. The children themselves are so busy with 'projects' and other work for examinations that they have less time or inclination for recreational reading. Reading stimulates the imagination by the creation of mental pictures from words on a page and as such it helps create those all important neural networks in the brain. But reading is too passive and too slow a pastime for many children today, indoctrinated by television and videos with their constantly changing images [see 173; 240; 285; 300]. Children who read fluently often show better performance in all their subjects. Fortunately, after a few decades when it was abandoned at the behest of educational theorists, the phonics system of reading is now back in primary schools (where children learn to read by learning the sound of individual letters). Learning a second language to read and speak broadens this skill even further. While learning to walk and talk comes naturally, learning to read and write are applied skills that require focused attention. The effort required by the child is itself an emotionally rewarding process and teaches them that prolonged effort may be needed to fulfill some of life's achievements – what psychologists call deferred or delayed gratification (see Chapter 2).

There is continual discussion as to whether the 'improvements' in results in school examinations in Britain have been achieved at the expense of standards. The standards of examinations, at least in mathematics and the sciences in Britain, with which I am familiar, do not seem to have changed substantially, either up or down over the past forty years that I have been teaching. There are certainly much simpler science courses available now and most students are opting for these. But it is the format of examinations that has changed. Pupils are now programmed much more thoroughly than those of yesteryear to know what kinds of examination questions they will face to increase the chance of success –

which is good for the student and for the school, but it gives the pupils 'tunnel vision' in the subjects concerned: they learn only what they need to learn for specific kinds of exam questions. What is lacking is the breadth of study and any time in classes for communicating the excitement of the subjects or the history of their development which, in science particularly, is a major loss.

It is regrettable that 'league tables' of school results have turned education into the sort of stressful competitive 'rat-race' of capitalist business, to the detriment of both teachers and children. The testing and examination results have now become the objective instead of the learning. The process of teaching needs to be more than just imparting factual knowledge to be regurgitated in examination questions. It should aim to stimulate pupils to ask questions of their own. Good teachers will always do this, if they have time and freedom to teach a subject in their own way. Pupils need to be encouraged to learn not in a mechanical way by rote (except for some boring but necessary tasks like multiplication tables!) but by being told (or, better still, to work out for themselves) the logical relations between factual data. This keeps a subject alive and exciting. The 'free associations' devised by Freud are examples of his kind of positive thinking. The 'peg' and 'link' methods of remembering data are also examples. In the 'peg' method, students remember new facts by associating them with other facts already well known, which serve as pegs on which to hang new data. In the 'link' method, strings of facts are memorized by finding a logical connection for each one with the next.

The detrimental effects of the sacrifice of physical education classes and games afternoons to cram in more academic work demanded to satisfy bureaucratic assessments were commented on in Chapter 1. This lack of exercise is undoubtedly one of the factors contributing to childhood obesity. Physical exercise in schools for growing bodies is even more essential as urban sprawl increases and grassy play areas become harder to find. Children's safety issues are other factors inhibiting what Sue Palmer calls the 'free-range child' [285, p.59], so school playtimes and exercise sessions become even more important. Herding children into large schools by buses, instead of their being able to walk or cycle to smaller, friendlier local schools, has exacerbated this problem. Once the oil runs out later this century we will be forced into thinking again about our philosophy of centralization and exploring a return to self-sufficient local communities of both food production and supply, and education. The time to start planning and executing this philosophy is now, rather than do what politicians always do – wait until the last minute and then hand difficult problems with unpopular solutions on to the next government, by which time they may be insoluble, or only soluble at much greater cost and difficulty.

Team sports also encourage social empathy and foster competition that is not subject to material rewards. What used to be known as adult 'sport' has now virtually disappeared within a generation. 'Sport' is defined as 'recreation, play, a pleasurable pastime' but capitalist economics has engulfed this too, so that what used to be sport is now simply competitive business conducted in the open air. It's good for children to have the opportunity to enjoy sport as play without the commercialism. The pleasure of play involving positive stress and stimulation is an essential component of happiness, for children and adults.

Religion

Freedom of expression and of worship are quintessential components of holistic living, so the proliferation of single-religion schools encouraging children to develop, overtly or implicitly, intolerance of every other spiritual world-view is surely not the way forward. It is immoral to corrupt young minds in this way, however noble the intent. Children need moral guidance but this does not have to be accompanied with religious indoctrination: there are universally agreed principles of morality. Many philosophers, like David Hume, G.W.F. Hegel, Adam Smith, Erich Fromm and Mary Midgley, have pointed out that social advancement can only occur when there is empathy between the various individuals in society and this cannot be achieved from only one religious standpoint in a multi-cultural society.

We should be grateful to the rationalism of science for having provided us with all the benefits of modern living and for its having replaced the fear and superstition of pre-Renaissance religious thought. But in the process science has also extinguished the sense of mystery, of awe and wonder at the grandeur of the cosmos. It is this spirituality, whether secular or religious, that we must recapture to give us our sense of communion with one another and with the Earth. This great mystical awareness blended with our rational scientific knowledge is the New Story that we must embrace instead of the dogma of religion. For social cohesion in a productive society, the dogma proclaiming the unilateral truth of each religion must be replaced by an all-embracing, unifying spirituality to wean our youth away from the 'spiritually desiccated consumer society out of touch with the numinous powers pervading each being in the universe' [360, p.35]. As Brian Swimme says, we need to develop 'humanity congruent with the ways of the universe' [360, p.45].

It is futile and retrogressive to suggest that we should campaign for the abolition of all religion because of the atrocities committed by fundamentalist extremists. For many people, the practice of the rituals of moderate religious belief sustains and nourishes their faith and, for them, it is indispensable. Indeed, as we have seen above, some form of worship of divine nature spirits seems to have accompanied humankind since the beginning of human development, but this simple pagan spirituality was crushed, most effectively, by the spread of Christianity.

It is not religion that is the 'root of all evil', as Oxford biologist Richard Dawkins maintained in his programmes on British television in 2005, but extremism and intolerance – of religion itself by atheists as well as of the views of other faiths by believers. Professor Dawkins says repeatedly in interviews that it is the 'certainty' of religious beliefs that worries him. Any rational person who has heard Professor Peter Atkins, from the Chemistry Department at Oxford, talking on religion will be equally disturbed at his 'certainty' at the foolishness of anyone who believes in an interactive spiritual domain, for which there is abundant scientific evidence – rigorously tested empirical observations. Spiritual world-views are simply the complement to the rational viewpoint of science, just as we have complementary and allopathic medicine. However, as Dawkins said, for children to be educated in single-faith schools is without question a backward step away from holism – though not because scientific rationalism was being replaced by primitive superstition as described by Professor Dawkins. As I pointed out in Chapter 4, the scriptural basis of religion is not 'superstition' but a tradition of myths and rituals handed down through many generations, or an interpretation of the spiritual insight of one prophet as to how we should live our lives. It is the indoctrination of people, and especially children, by clerics of the various religions who present these scriptural myths as absolute truth that is deceitful and evil. Where orthodox religion was crushed within some secular cultures in the 20th century, we had instead a political ideology as religion, and the humanist record of these cultures is appalling: crushing religion certainly did not produce a utopia! The evil of single-faith religious indoctrination is the real or implied intolerance of other world-views of which children especially should be aware.

As part of a change in outlook of humankind it is more important that children are taught about the undivided wholeness of cosmic spirituality that permeates the entire natural world rather than be indoctrinated with rigorous and inflexible man-made dogma from specific religions. Such spirituality should be uppermost in all world religions, for it is within each of us. Children must be made aware of the true nature of the scriptures and the religions based on them if we are to create an interactive holistic world – which surely we must if we are to survive.

History

I have never taught history, but from the material I have seen in schools and in pupils' text and exercise books, there seems to be an undue focus on wars and treaties in the history syllabus. It would be more constructive in history lessons to focus on the achievements of humankind over the past two millennia, or even five millennia if time permits, instead of subjects like World War II, which seems to have become something of an obsession with history syllabuses in Britain. This is reflected by the content of the UK History channel on digital television. That is not to say that children should not be made aware of the depths of depravity to which some powerful figures have reached at many points in history; but the focus must surely be on the positive achievements. History syllabuses in America tend to centre on the four hundred years of the development of the United States, as I understand it from friends there.

In Britain or America, there seems to be little focus in history teaching on the huge constructive social revolutions of the past millennia – the agrarian revolution, the Renaissance, the age of discovery and exploration, and the Industrial Revolution with their benefits (and many less desirable consequences), the re-invention of printing in the West, and similar topics of human achievement instead of endless lessons on wars and battles that emphasize the aggressive power-driven quality of human nature. At a time when the old hatreds of the Crusades and Inquisitions between Christianity and Islam are having such appalling consequences on the way we live our lives, why do we not hear about the culture of medieval Islamic al-Andalus that achieved cooperation between Christian, Jewish and Islamic scholars to diffuse the learning from ancient Greece and Rome to the rest of the world? To focus on the evil that men do does not implant in the unconscious of young minds a loving, tolerant, holistic attitude to life for the adult.

Mathematics

Mathematics is a holistic symbolic language. It crosses any linguistic barrier between nations. It can also be applied universally in any subject area, though new terms might be introduced to deal with specific concepts in a particular subject. Once the vocabulary of symbols and the grammar of operations have been mastered, as we must do with any language, there is a timeless universality about the relationships it expresses. Again, as with any other language, skill develops only with repeated practice. One hugely constructive property of mathematics for schoolchildren is that, in its problems, it presents the prospect of reward (through correct answers) within a few seconds of effort. There are few challenges in life where, within the space of a few minutes, one can subject oneself to repeated challenges and get the satisfaction of continual successes, provided of course one has made the initial effort to grasp the necessary techniques.

Apart from its being the language of nature and therefore of science, mathematics has a beauty of its own and there has always been a particular affinity between music and mathematics. The Pythagoreans of ancient Greece were aware that only when the vibrational frequencies of sounds were in a whole number ratio were the sounds pleasing to the ear: fractional ratios sound dissonant.[2]

An appreciation of the logic of mathematical reasoning is a basic training in logical thought of any kind. The geometry of Euclid's *Elements* was taught in British schools until the middle of the 20th century. Then, in response to a demand to make school curricula 'simpler' so that they were accessible to all, the teaching of most of this formal geometry, as well as much algebra, was dispensed with and, as a result, this valuable instruction in logical thought was lost. Indeed, too often what is left of geometric theorems seems to be taught as just more factual learning instead of as a dimension of logical thought. Algebra becomes nothing more than a tedious set of rules for solving problems instead of a generalization and extension of arithmetic operations.[3]

When, in the 12th and 13th centuries, Christians discovered the wealth of scholarship in the newly conquered lands of al-Andalus, much of it was unintelligible to them, being written in Arabic. They needed scholars from the rest of Europe who had studied Arabic to help them decipher the writings of the ancient Greek philosophers and even that of some of the early theologians that had been preserved so carefully by Islamic, Jewish and Christian scholars for hundreds of years. One such scholar-translator was Leonardo Fibonacci of Pisa (*ca.*1170-1230). He discovered a sequence of numbers that has many practical associations in Nature and in human art and architecture[4].

The Greek philosopher Plato and his pupil Aristotle thought that five solid shapes – the cube, icosahedron, octahedron, tetrahedron and dodecahedron (called the perfect solids) – were the building blocks of matter, because they are the only solid shapes to have identical faces made up of regular polygons. They therefore designated these five Platonic solids as representing the five elements – earth, water, air, fire and ether – corresponding to the solid, liquid, gas, plasma and energy states of modern physics. The measurements made by the Greeks, and by the Babylonians before them, together with the manipulations of those measures that they invented, enabled humankind for the first time to quantify some of the dimensions of our cosmos that had been a source of awe, wonder and mystery since Man's first appearance on Earth.

Telling stories such as these to pupils, and exploring some of the fascinating properties of numbers[5], would not be of any use in passing modern examinations in maths. But how much more exciting and relevant the subject seems when abstract numbers can be shown to be a function of the natural world and to find application in some of the great constructions of ancient and modern architects.

Physical sciences

It is judged that there is no longer room for the history of science in science syllabuses in Britain. So most students will now leave school with little or no awareness of the contributions to human knowledge of Copernicus, Galileo, Newton, Darwin, Lyell, Pasteur or Einstein. Few schoolchildren today will ever have heard of Priestley or Lavoisier, phlogiston or caloric; the experiments of Rumford that helped unite the concepts of mechanical work and heat; or Faraday and Maxwell's unification of electricity and magnetism; and their education is all the poorer for it. These were the first steps taken in the 19th century to unite the fragmented subjects of scientific study. The history of science is intimately bound up with the philosophy of science.

There seems to be no treatment in schools of what science really means – observations of the natural world leading to hypotheses; a testing of those ideas by carefully chosen experiments; verification or falsification of the ideas leading, if necessary, to new hypotheses that with continual testing may become theories or established laws of Nature. Science is not primarily a body of 'facts', though it seems to be taught as such, but an ever-changing assemblage of world-views on different subjects. If a theory can be supported by the logic of mathematics then so much the better, but science is as much a work of imagination and inspiration as it is of practical experimentation and calculation [401, p.101]. Even scientists can conceive only that which comes to them through infusion of the universal spiritual energy of the cosmos: as Brian Swimme says, 'To fully understand, one has to sit down and wait for the universe to enter' [360, p.29].

The most modern currently accepted scientific world-view is now generally called a paradigm [208]. Sometimes however even the basic paradigm in a particular area is shattered. The Greek philosophers Leucippus and Democritus living in the 5th century BCE believed that all matter was made up of indivisible and indestructible small particles they called atoms. The idea was formalized two thousand years later by John Dalton's Atomic Theory. Yet before the end of the 19th century our ideas had to be revised when Henri Becquerel discovered radioactivity. The idea that light traveled in waves was the predominant view since the beginning of systematic scientific experimentation, so it was quite revolutionary to find evidence that it also behaved like a stream of particles. Albert Einstein was awarded the Nobel Prize for his experiments on this photoelectric effect, though Newton had suggested a corpuscular theory of light three centuries earlier. The behaviour of matter according to the principles established by Isaac Newton seemed well established for three hundred years until the mathematical and experimental work of the quantum physicists showed that very tiny particles behaved in ways that not only defied Newtonian physics but even the laws of common sense. Perhaps the real world is quite different from our description of it, and even that will change as we make more observations and measurements. The very cause-and-effect principle on which the whole of scientific induction is based has been called into question by some philosophers, like David Hume and Karl Popper. The moral of our story is that students should be aware that science gives us a world-view based on our five senses and reason; it is temporary at a very fundamental level and is constantly subject to revision, but it works in practice. Mysticism on the other hand gives us a world-view based on our sixth sense but one that claims to present eternal truths.

The scientist schooled in logical positivism will say that science is about ideas that can be subjected to rigorous testing that will either prove or disprove any conjecture. Any notions not amenable to such testing are pseudo-science or fantasy. The results of such tests must be totally consistent at all times and places. Yet there are few who would not consider Darwin's theory of evolution as science: but we cannot reverse time and prove that his proposed evolutionary pathways are true, only that they (mostly) fit the observed facts. The same is true in geology. There is a principle called Uniformitarianism that *assumes* that 'the present is the key to the past'. In other words, the Earth processes we see today are *assumed* to be essentially the same as those that took place throughout Earth's geological history.

Cosmologists must speculate about the origins of the universe, for this was a one-time event as far as we know and it cannot be investigated in the usual scientific manner. Scientific techniques have served us well; but not everything in science can be subjected to 'double-blind' testing (where experimenters do not know the identity of the subjects or materials under test). When we enter the world of biology or sociology, where animal or human subjects are being examined, the criterion of invariance simply cannot be adhered to rigorously. Even the same person subjected to the same test at different times might behave differently. This is another important principle of science that students should be aware of.

Few pupils will ever have encountered the terms 'reductionism' and 'holism'. The holistic world-view we seek to teach our children today has evolved from the wealth of knowledge that we have acquired through reductionism over the last four centuries. There is no suggestion that we should abandon the reductionist approach in our investigations. But the acquisition of knowledge is an evolutionary process, and we will stunt the development of children and of humankind in general, and possibly lead to its extinction, if we continue to believe that the reductionist approach is the *only* one that will provide us with knowledge – learning more and more about less and less. We *must* see the global picture. We *must* accept that there is a spiritual non-empirical dimension to the world that gives us insights into our past mistakes and to our future path for the holistic unity of humankind. This is one of the most important lessons that science can teach our children – that it is not the only way of viewing the world.

Biology

One of the most important biological issues for adolescents will undoubtedly be that of sexuality. Although in adolescence sexual attraction is often the driving force for establishing relationships, it is important that teenagers express their emerging sexuality responsibly, both for their own wellbeing and for others. World population continues to grow exponentially. Knowledge of birth control methods is now readily available, yet young teenage girls are still becoming pregnant before they are physically or emotionally ready for motherhood. In the West, where Christianity has dominated religious thought for many generations, human sexuality has been regarded, at best, as a necessary evil for procreation and at worst as an example of human sinfulness in indulging in perversion or depravity. Erotic literature of all kinds has come to be labeled as 'pornography'; but there is nothing intrinsically corrupting about the expression of human sexuality in art or literature. Giving sex instruction in schools purely in biological terms ignores the much more important role of sex in human loving. Sex has become a commodity like so much else in our consumerist society, a superficial and transient pleasure that is bought and sold, and we should not be surprised at the resulting increase in prevalence of sexually transmitted diseases. It is time that sexuality was talked about openly, in biology classes or elsewhere, in its emotional and social context and we abandoned these highly detrimental man-made religious proscriptions on birth control.

One of the key issues in biology currently is the controversy about the teaching of Darwin's theory of evolution in relation to the biblical account of Creation. This discussion is once again prominent in America and, no doubt, soon will be in Britain. The term 'Intelligent Design' (ID) is used currently to denote the belief that the evolutionary process has not proceeded through blind chance as the majority of biologists claim but that there has always been a purpose or goal of human evolution or, to put it another way, that there is a Guiding Hand steering the process along in some way. The ID argument gets its strength from at least three sources.

First, without some 'designing' force, it is statistically improbable that simple molecules could assemble themselves into the complex structures of life forms, however long they had available, and whatever intrinsic potential they may have had for assembling into complex structures. There is in biological structures an 'irreducible complexity' [22; 375]. Put another way: it doesn't matter how many monkeys we have working on an array for typewriters for an unlimited time, they will *never* produce the works of Shakespeare or Einstein's theory of relativity because there is an irreducible complexity in the human brain that far exceeds anything that has so far evolved in other animals. Current theories of the emergence of complexity as an inherent property of molecules and their subatomic constituents do not invalidate this argument. The idea of an all-pervading spiritual energy – the quantum or zero point field – that some would want to 'personify' as God is still rationally tenable as a 'creating' and 'designing' agency.

Secondly, the basic building blocks of the molecules of life show a 'handedness' or chirality as the chemists say. The carbohydrates that form the food and structure of plants are polymers made of sugars; animal bodies are made of many different varieties of protein polymers and these are made up of some twenty amino acids. Most natural sugars and amino acids show this property of favouring one kind of handedness. If the left- and right-handed kinds were randomly distributed in Nature, as they usually are in the products of chemical reactions in the laboratory, stable carbohydrates and proteins could not have been formed.

Thirdly, there are many facts about the material properties of the universe that make it uniquely suitable for human life. If there were more oxygen in the atmosphere, lightning strikes would produce massive incineration. If there were less oxygen, we might have needed gills instead of lungs. Water is everywhere on Earth in our clouds, rivers and seas and we saw in the preceding chapter how it comprises one of Earth's natural cycles. But water is both a conductor of electricity and an insulator. It is a solvent for many compounds. It is neither acidic nor alkaline but neutral. Its maximum density is not at its freezing point, as with most other liquids, but at 4°C. If the so-called fundamental constants of Nature were just very slightly different, life could not have evolved. There are so many properties of our world that are perfectly suited to Man – an idea known as the Anthropic Principle [20; 47]. The idea that humankind is a part of some cosmic spiritual energy is a much more holistic concept than regarding human existence as without meaning and arising by chance.

The Darwinism–Creationism argument has implications that go beyond the specific issue of evolution. What is at stake really is the question of whether the reductionist and materialist methods of science are adequate to explain all natural phenomena and whether biblical stories can provide any meaningful information regarding natural phenomena. The hyperbole of some creationists in describing Darwin's theory and the flat-earth theory as being of equal credibility is just nonsensical and unhelpful.

We need to be clear about exactly what we mean by Creationism. One interpretation is that it means belief in the literal truth of the account of Creation given in Genesis, which is what most Christian fundamentalists believe. Supporters of this viewpoint then go on to conclude from biblical references that the Earth is some 6000 years old – a conclusion that directly contradicts much unequivocal scientific evidence from geology and radiocarbon dating techniques. These and other calculations give us an age for the universe of some 13 to 14 billion years and for the Earth of about 4.6 billion years. The other meaning attached to Creationism is Intelligent Design – belief that some divine spiritual force, energy or Being initiated the act of Creation and subsequent evolution according to the principles outlined by Darwin. Even most theologians today regard the Genesis account as mythical or allegorical and it certainly has no place in any scientific description of the origins of the universe and of life on Earth: 'Creationism is both scientifically untenable and theologically misguided' [see discussion in 97, pp.49-55; see also ref.9]. There are enough unresolved issues with Darwinism to occupy rational debate in science classes without introducing wholly inappropriate discussion of scripture.

The important point to get across to pupils is that Darwin's theory is just that – a theory. It is not incontrovertible fact like the law of gravitation. There seems little doubt that it is essentially correct but major points remain unresolved, the most damaging of which is considered to be the absence of intermediate life forms in the fossil record. Perhaps one could say that Darwinism is necessary but not sufficient, to use the terminology of the mathematicians. That does not mean however that the Genesis account of Creation should be taught in sciences classes alongside Darwinian theory. Darwin's account is science, based on empirical observation and consistent with a mass of other evidence from biology and geology, even if it cannot be tested and proven to order, like much in science. Most theologians agree that the biblical account of

Creation on the other hand is myth or allegory woven out of historical people, places and events.

Another aspect of biology teaching that could be given a holistic slant is cell biology. The earliest life forms were single-celled prokaryotes, which do not possess a nucleus. Bacteria and cyanobacteria are the only prokaryotes alive today. Eventually these prokaryotes evolved into much more complex kinds of cells called eukaryotes. These cells demanded cooperation between their different components for their survival. The functions of eukaryotic cells are miniature versions of the functions that take place in whole organisms, which are made up of organs that are made of tissues that are made of cells – remember the holon concept we met in the Introduction. Eukaryotes also exhibit 'intent' and 'purpose' in that they actively seek environments that will support them and avoid toxic or hostile ones that would harm or kill them. In fact, as pointed out by Bruce Lipton [221; 222, p.37], 'there is not one 'new' function in our bodies that is not already expressed in the single cell'. In David Bohm's terminology [31], the human organism represents one aspect of the implicate order of the universe. Eukaryotic cells cooperate with one another within the organism, but their component organelles must also cooperate within each cell. We know this today, but even two centuries ago the French biologist Jean-Baptiste de Lamarck (1744-1829) said in his *Philosophie Zoologique* (1809) that whole organisms evolved successfully by cooperation. This is a view endorsed by British physician Frank Ryan, who lists a number of such examples in his recent book [322, p.16]. The importance of cooperation between individuals for the wellbeing of society rather than the Darwinian notion of competition has also been emphasized by philosophers like Erich Fromm [112] and Mary Midgley [253], as mentioned earlier.

There are many who believe that it is Lamarck, rather than Darwin, who should be regarded as the founder of the theory of evolution [243, p.227]. Lamarck believed that organisms were affected by their environment and could pass on resulting influences to their offspring.

Commenting on Lamarck's views, Darwin wrote in the Historical Sketch introduction to his *Origin of Species*: 'With respect to the means of modification, he attributed something to the direct action of the physical conditions of life, something to the crossing of already existing forms, and much to use and disuse, that is, to the effects of habit'. Modern cell biologists believe that the cellular environment is all important to the expression of gene activity. So important has this subject become that it has become a recognized research field called *epigenetics*. The environment does not change the primary gene sequence but rather modulates when and how genes are expressed. Genes are not 'self-emergent' – something in their cellular environment has to trigger their expression of proteins. This is how exposure to the electric fields produced by overhead cables and our mass of domestic electrical equipment may produce ill-health in susceptible individuals, as reported by some researchers [274]. It is also how perpetuating the emotion of love instead of hate benefits us as individuals and as a global society by inter-action through the collective unconscious and our own genetic makeup.

The message for pupils is that the bodily environment directly affects the function of the cells that comprise all of our tissues. Any toxic drugs, including tobacco and alcohol, will therefore have a direct and adverse effect potentially on every cell of the body. It is therefore vital for the wellbeing of a foetus that pregnant girls should not drink or smoke *at all* during pregnancy. Individuals will prosper only in an environment that is not just benign but uplifting. This applies to one's peers and to the whole school ethos. Boys and girls need to be surrounded by other pupils with a positive work ethic, or at least are tolerant of those who do, eradicating bullying whenever it occurs. Schools need to provide nutritional meals and disciplined classes in which pupils have an optimal learning environment. Only if we provide the opportunity for the younger gener-ation to take on this positive philosophy from home and school do we have a realistic prospect of humankind thriving into the 22nd century.

Cosmology

There is no Cosmology in school syllabuses so, for the present, this will have to be fitted in elsewhere. I have talked of these subjects as separate topics, as they are taught in most schools, but a holistic approach would be to apply a more integrated educational philosophy like that in the Rudolf Steiner schools.

At the dawn of science in the 16th and 17th centuries, the whole world-view changed with the Copernican theory of Earth revolving around the sun instead of being at the centre of the universe. The studies of Galileo and Newton confirmed that the old Ptolemaic geocentric theory approved by the Church had to be abandoned. This is such a commonplace idea to us today we cannot begin to experience the impact this idea must have had four centuries ago on people who saw the sun rise and set 'around' the Earth, just like our moon. Yet it was we on Earth who traveled around the sun. This was not just another scientific observation – sun and moon 'moved' as they had always done; but the interpretation, the vision of what these phenomena meant led to a different world-view. In the 20th century, with the theories of Albert Einstein and the observations of Edwin Hubble, our cosmos is seen as everywhere expanding outwards from any viewing position within the galaxy. So Earth has once again become the centre point of reference for the cosmos. The universe is vastly more populated with stars and galaxies than Copernicus could ever have dreamed of. It is this vision of awe and wonder of our place in the cosmos that we must communicate enthusiastically to the young generation to give them a sense of grounding and belonging within the cosmos.

Summary

The future lies with our children and grandchildren. We as adults have to behave in a way that sets the right example for our children. We need to look after our bodies and eat sensibly, taking as much enjoyable exercise as we can, preferably with the children (walking, cycling, swimming). Drugs of any form, including tobacco and alcohol, need to be avoided, and naturopathic rather than allopathic remedies used for ailments as far as possible. We have to work in order to provide necessities for ourselves and for our families. Work also provides us with a sense of satisfaction and a chance for social contacts. But work is only part of our daily schedule. The joy of holistic living should be our goal, not material acquisitions.

The direction our life takes depends to a large extent on the decisions we make. Children must be dissuaded from a hopeless fatalistic outlook that there is nothing they can do to change what will happen to them. Our genetic programming and childhood nurture certainly lay the ground for what happens to us as adults. But still, once we reach maturity, we have freedom of action. We do not have to follow the herd if we feel there is a better Way. With determination we can overcome many of the obstacles with which life will almost certainly present us. Our thoughts and actions will determine our life path and, to some degree, that of others.

Schoolchildren are slowly learning to care about the environment. Indeed, some primary school pupils seem to have greater awareness of the problems than many adults! Whilst this chapter has focused on education in schools, the first part of the education process takes place in the home. For this, adults themselves need to be convinced that tolerance and an interactive approach to life is essential if our children are to become better caretakers of the world they inherit. This attitude is what we need to build on so that today's children can live holistically and inherit a stable, loving, bountiful Earth. There have been at least five mass extinctions of life on Earth since it began. We do not want our children to witness the sixth extinction brought about by Man [218].

NOTES

Chapter 1

1. The BMI is calculated by a simple sum: weight in kilograms divided by height in metres, squared. The resulting values should be in the range of 20-25. If the calculated result is less than 20, this suggests you are underweight. BMI values in the range 25-30 tell you that you are overweight, while obesity is classified as a BMI of more than 30.

2. Cycling, swimming, playing any kind of sport, gymnastic workouts, or gardening and walking if one is aiming for the gentler type of exercise, should benefit every body. Jogging should always be undertaken in rural areas, countryside, parkland, or on the beach where the air is purest and pollution least. Pounding unforgiving pavements in heavy city traffic is likely to do far more harm by damage to the spine and inhalation of toxic motor exhaust than any possible benefit to the cardiovascular system. What we can enjoy or physically tolerate by way of exercise or sport will vary from person to person, but walking, gardening, cycling, or swimming are activities we can follow at our own pace, and seem to produce mental uplift for most people. For those who are forced into a more sedentary lifestyle there is always the mental stimulation of chess and other board games, and crossword and puzzle books with challenges like sudoku, to keep the mind active.

3. In 1977, the Hyperactive Children's Support Group was formed in Britain to offer help to parents (Director: Sally Bunday; contact at 71 Whyke Lane, Chichester, West Sussex PO19 7PD; tel: 01243 539966). In 1978 Amelia Nathan Hill set up an organization in Britain, Action Against Allergy, that people could turn to for advice if they suspected they or their children were suffering from 'allergies' (Contact: PO Box 278, Twickenham TW1 4QQ; tel: 020 8892 2711). Colourings are usually in the E100s. The benzoates (210-219) and sulphites (220-227) used as preservatives are particularly bad in producing allergic reactions in

children. The flavour enhancer monosodium glutamate (621), which was originally isolated from seaweed in Japan, produces headaches and bloating in many people.

Richard Mackarness' pioneering studies on hypersensitivity of thirty years ago have been confirmed by subsequent research. Professor John Warner at the Department of Child Health facility located at Southampton University in Britain is conducting ongoing studies that are published in the *Pediatric Allergy and Immunology* journal, of which he is editor-in-chief, and elsewhere.

4. There are five sweeteners authorized for use in Britain: saccharin, aspartame (NutraSweet, Canderel; E951) for which there are many reports of adverse effects from high quantities; cyclamate (Sucaryl, Sweet'NLow) which may be carcinogenic in high quantities and the use of which is banned in the US; acesulfame K (Sunette, SweetOne), possibly carcinogenic in high doses; and sucralose (Splenda), a trichlorinated sucrose for which there are no adverse effects reported but which hasn't yet been extensively tested. The chlorine in sucralose is *not* just like the chlorine in common salt (sodium chloride) as claimed by the manufacturers as the nature of the chemical bonding is different.

5. This work was based on the results of the studies by the Medical Research Council carried out by a doctor, R.A. McCance and a nutritionist, E.M. Widdowson over a fifty-one year period, 1940-1991.

6. It has been reported quite recently that the water companies in Britain spend £16 million each year to remove nitrates from drinking water and £120 million annually to remove pesticides [303].

7. Pesticide residues were found in 56 out of 72 samples (78%) of bread samples tested by the UK government's Pesticide Residues Committee in 2003.

8. Antimony citrate complexes have been found in fruit juices stored in bottles made of PET (polyethylene terephthalate) [145; 336]. Antimony trichloride is used as a catalyst in the manufacture of PET. Because it is a suspected carcinogen, the European Union designates it as a priority

pollutant.

9. The term 'allopathic' was introduced by the founder of homeopathy, Samuel Hahnemann.

10. The origins of pharmacology go back to the mid-19th century with the experimental studies of the effects of drugs (on animals) of Rudolph Buchheim and Oswald Schmiederberg in Germany, and the Pharmaceutical Society of Britain was founded in 1841 (the prefix 'Royal' was added in 1988). However, the present-day pharmaceutical industry gathered momentum after the discovery of a synthetic route for the preparation of aspirin by Hermann Dreser in 1893. An extract of oil of wintergreen (*Gaultheria procumbens*), a common shrub of the heather family found in Britain and North America was used as a traditional herbal medicine for relief of rheumatic pain long before the active ingredient, or its close relative, aspirin, was synthesized.

11. For example, there are some powerful allopathic drugs (like finasteride) available to treat benign prostatic hyperplasia (BPH – an enlarged prostate gland leading to difficulty with urination) that is common in men in later life, but all of them frequently produce some rather nasty side-effects. The gentle natural compound saw palmetto (an extract of *Serenoa repens*) is just as effective with, generally, only mild or no side-effects [28(a)]. Readers should note however that orthodox medical opinion is sharply divided on the treatment of BPH because such treatment may mask prostatic cancer. However, there is no reliable method of assessing prostate cancer, and the two basic treatments – surgery or radiotherapy – frequently result in impotence or incontinence or both, with no convincing evidence that such radical measures prolong life [28(b)].

The family of drugs called statins is very popular with doctors at present for reducing cholesterol levels in the blood and, provided the patient is able to metabolize them, they are quite effective. Statins work by inhibiting the production of cholesterol in the liver, most of all at night, but they also inhibit one of the body's key metabolic agents, co-enzyme Q10. Statins frequently produce insomnia, myalgia (muscle pain in the legs or arms), and shortness of breath as side effects and can even damage muscle tissue (myopathy) [210(a)]. When the patent was first granted for the use of statins it was recommended that doctors prescribe co-Q10 at the same time, to minimize their adverse effects [210(b)], but this seems to be done only rarely. There is a gentle herbal alternative to statins available and that is to use spreads, yoghurts or tablet supplements containing plant sterols and stanols. In the human diet they appear to lower the undesirable LDL cholesterol while leaving the desirable HDL cholesterol unaffected [55; 215; 271; 292]. Antioxidants like vitamin C, vitamin E, the flavonoids and xanthones also seem to have a beneficial effect in inhibiting atherosclerotic plaque formation. Vitamin C is to be found in all fresh fruit and vegetables. You can find flavonoids in citrus fruit, vegetables (especially onions), legumes and green tea, or even in capsule form if necessary. Vitamin E is obtained from wheat germ and sunflower seeds and mangosteen is the best source of xanthones.

While cholesterol levels may well be an important factor aggravating arteriosclerosis (blocked blood vessels), there are other compounds, like homocysteine, the blood concentration of which is believed to be even more important [244], but this is never measured routinely. High protein diets, especially those that include much red meat, tend to produce elevated homocysteine levels, which are known to be detrimental to health.

The usual treatment prescribed for osteoarthritis is to take pain-killers. As Lynne McTaggart says in her book: 'Conventional medicine tends to take the view that there is no known cause or cure for arthritis, so all it can do with certainty is to alleviate your pain' [248, p.206]. These drugs all have damaging effects. Paracetamol can be taken in only limited quantities because of its toxic effect on the liver. The non-steroidal anti-inflammatory drugs (NSAIDs) all produce irritation of the gut, and may cause difficulty in breathing in asthmatics. The American FDA estimates that there may be up to a quarter of a million cases of gastric bleeding from use of NSAIDs each year, while in Britain about 4000 people die from taking NSAIDs each year. Again, there is a naturopathic alternative. The natural compound glucosamine, taken with fish oils containing omega-3 fatty acids, may prove to be a safe and simple alternative in the treatment of osteoarthritis [310].

There have been quite extensive clinical trials on humans for the past decade or so on the use of extracts of *Ginkgo biloba* for the treatment of dementia, Alzheimer's disease and poor circulation [28(a); 219; 349]. With fewer clinical trials so far, the xanthones from mangosteen have also proved to be effective. The staff at the Human Nutrition Research Center on Aging at Tufts University, Boston, MA, have conducted research into mangosteen over the past decade [306]. The results have been positive and side effects are generally few – mild headaches or gastrointestinal upsets. This is another example where a gentle herbal remedy may obviate the need to use more toxic synthetic pharmacological compounds.

12. Pollen of various types, house dust mites, mould spores and animal dander are common allergens, as are foodstuffs like eggs, milk and dairy products, wheat, potato, chocolate and peanuts. The allergic response produced is usually difficulty in breathing, uncontrollable sneezing, or skin rash, but headaches or a bloated tummy are other common symptoms.

13. The seven *chakras* are associated respectively with the brain stem and pineal gland (crown *chakra* that facilitates meditation and spiritual communion), pituitary (the third eye that enables us to retain a balanced spiritual/material world-view; enhances psychic ability and allows the 'being' mode of Fromm's outlook to dominate the 'having' mode – see Chapter 2), thyroid and parathyroids (throat *chakra* that facilitates our communion with others), thymus (heart *chakra* generating universal love), adrenals and pancreatic glands (the *chakra* associated with the solar plexus, governing will, intention and self-control), sex organs (the *chakra* representing desire, sensuality and sexuality), and the base of the spinal cord (the base *chakra* associated with the ego, the sense of self and empowerment). The crown *chakra* connects with the spiritual domain

while the base *chakra* keeps the individual grounded. These chakral assignments are spiritual in nature and have little to do with the physiological functions of the organs. That said, in everyday speech we all know that if we experience emotional distress we feel 'heartbroken' even if emotions are attributed by physiologists to brain function. Individuals who have a low feeling of self-worth often experience low back pain. Feeling the pain of others gives us 'a lump in the throat'. So we have traditionally associated certain emotions with these chakral functions.

14. The ancient Greek philosopher Aristotle thought that happiness was the real goal of human existence as an end in itself, rather than an intermediate to something else. The 19th century philosopher John Stuart Mill (1806-1873) thought that the quest for happiness should define our standard of moral action: 'actions are right in proportion as they tend to promote happiness, wrong as they tend to produce the reverse of happiness. By happiness is intended pleasure, and the absence of pain; by unhappiness, pain, and the privation of pleasure' [258, Chap.2]. Most ethicists today would have some reservations about defining morality on the basis of utilitarianism, and psychologists recognize another dimension to happiness besides pleasure and pain and that is satisfaction or fulfillment.

15. Myths and traditions are an essential component of the spiritual fabric of society, as described in Chapter 4, transmitting ideas through the ages from our ancestors. The publication of folk tales became a fascination for writers in the Romantic era and many were set to music. One reason for the rise of Romanticism was the expression by the human heart and mind of an emotionalism that seemed to have burnt out in the furnaces of the Industrial Revolution. The interest in folk music in 19th century Europe was part of this expression of a search for ancestral roots. Perhaps the most distinctive characteristic of English music of the 20th century was its pastoral atmosphere portrayed by strings and woodwind, reflecting the idyllic English countryside. The lushness of much English music of the 20th century might also be regarded as a rebellion against the

urbanization and mechanistic preoccupations of Victorian England. In 1810 only 20% of the English population lived in towns, so most lived a rural life, but by 1910 this had risen to 80%. Russian music of the 20th century by contrast uses much percussion and heavy brass portraying, perhaps unconsciously, the brutality and impersonality of the Communist regime.

16. Deryck Cooke in his book *The Language of Music* [60] described music as ‘the most articulate language of the unconscious . . . the expression of man’s deepest self’. Cooke believed that music reflected qualities of other arts – of architecture in its formal pseudo-mathematical structure, of literature in its expression of emotion, and of painting in the representation of physical objects. These qualities are reflected in music as it evolved from the medieval period to the present-day. The music of the medieval, baroque and classical periods prized formal structure: sonata form, the string quartet and the symphony all developed during the 18th century classical period with the compositions of Joseph Haydn and W.A. Mozart. The Romantic period in music, as in literature, focused on the expression of emotion. The use of minor rather than major keys became a feature of music intended to convey sadness or nostalgia. The Impressionists in music, like Claude Debussy and Maurice Ravel, were so-called because their output generated the same kind of aesthetic atmosphere as their artistic counterparts – a nebulous mysticism.

In the 20th century, with the arrival of serialism, structure or form again became dominant, but the emotionalism expressed in the previous century in music refused to die out. The need of individuals in society to try to escape professional pressures and the over-abundant materialist philosophy that surrounded them produced so-called minimalist music in the later 20th century from composers like Philip Glass and John Adams. Some 20th century music, by composers such as John Tavener and Arvo Pärt, focused on expressing the spiritual dimension and has the same kind of ethereal quality created by medieval plainsong. Such music reflects the global expansion of spirituality over recent decades.

Chapter 2

1. Wilhelm Max Wundt (1832-1920) opened a laboratory for the study of psychology at Leipzig University in 1879. After graduation, Wundt obtained a post as assistant to the physicist Hermann Helmholz (1821-1894) in Heidelberg, subsequently becoming an assistant professor there in physiology. From 1875 to his retirement in 1917 Wundt was professor of philosophy at Leipzig. It was Helmholz who first suggested interference from the unconscious mind in conscious thought processing, though his name is best known to scientists as the man who showed that different forms of energy were equivalent. It was he who stated what became known as the First Law of Thermodynamics or Law of Conservation of Energy.

2. Psychology in the first half century of its existence as a separate discipline was what is now known as *cognitive psychology*, a study of how the mind works and how we gather and use information (sensory data). It was also mostly clinical psychology, dealing with actual patients in some state of mental distress. Those who tried to help patients clinically established the separate discipline of psychiatry: broadly speaking, psychologists are scientists, psychiatrists are doctors specializing in mental disorders. During the first half of the 20th century, psychologists realized that their observations and diagnoses of mind states were highly subjective and really known only to the subjects themselves. They wanted to be recognized as scientists, so they turned their attention to something that could be objectively measured – behaviour. The contents of the mind could never be known to anyone other than the individual patient except when it was demonstrated and inferred through their behaviour. So the study of mind became essentially *behavioural psychology*. In 1913, American psychologist John Broadus Watson (1878-1958) even went so far as to define psychology as the study of observable behaviour. Watson was followed in this belief by Burrhus Frederic Skinner (1904-1990), who always maintained that psychology could never be a science of mind for this reason. However, it is quite easy for someone's behaviour to

deceive an observer, as in the acting profession, so this is not a reliable assessment of the contents of mind. The areas of animal psychology and child or developmental psychology followed later and some other psychologists, like Freud, went down a different avenue to the behaviourists.

3. Sigismund Schlomo Freud was born to a Jewish family on 6 May 1856 in Freiberg, Moravia (now Pribor in the Czech Republic). Four years later (1860) the family moved to Vienna where Freud lived for most of his life. Freud was the first of 8 children and his father was a wool merchant. Freud graduated in medicine from the University of Vienna in 1881, age 25. After a few years of general medical practice, from the late 1880s he began to specialize in mental disorders and in so doing founded the practice of *psychoanalysis*. His first book, *On Aphasia*, appeared in 1891. Perhaps his most famous book, certainly among the general public, was *The Interpretation of Dreams*, which appeared in 1900. His method of psychoanalysis, still largely used today, involved analysis of dreams and word association. Freud died in London on 23 Sept 1939.

4. Alfred Adler (1870-1937) was also Jewish and was born on 7 Feb 1870 in Vienna, the second of six children. His father was a grain merchant. He graduated with an M.D. degree from the University of Vienna in 1895, 14 years after Freud, whom he first met in 1899. Thereafter he became a member of Freud's circle of psychoanalysts until 1911 when the two fell out over their radically different approach to psychoanalysis. Adler was not a psychoanalyst but an academic lecturer in psychology. On the 28 May 1937, he died suddenly from a heart attack, age 67, while on a lecture tour in Aberdeen. Adler's principal publications were *The Neurotic Character*, 1912; and *Understanding Human Nature*, 1927.

5. Erich Fromm was born in 1900 in Frankfurt, Germany and followed the same line as Adler in his psychology – that the development of human personality was highly dependent on social relationships. He completed his Ph.D. in sociology at the University of Heidelberg in 1922 but left

Germany in 1932 because he and his family were orthodox Jews and he settled in New York a couple of years later. Erich Fromm died in Switzerland in 1980.

Chapter 3

1. The original School of Buddhism is the *Theravada*, the 'Doctrine of the Elders', based on the direct teaching of the Buddha set down in the Pali Canon (*Tipitaka*). This teaching predominates in Burma, Sri Lanka, Cambodia, Laos and Thailand (Southern Asia). A later School, the *Mahayana* or 'Great Vehicle', accepts as scripture some later writings (*sutras* (Sanskr.) or *suttas* (Pali)), also incorporating sayings of the Buddha Sakyamuni, but containing more philosophy and poetic expression. This School predominates in Northern Asia – Tibet, China, Japan and Korea. These divisions are comparable to Catholicism and Protestantism in Christianity, or the Sunni and Shi'a in Islam representing different world-views but with the same religious figurehead. The core belief in *Mahayana* Buddhism is compassion for all sentient beings (the Great Compassion, *ahimsa*), the *bodhisattva* delaying his own entry to *nirvana* until he has done all he can to liberate other sentient beings, while *Theravada* focuses on the path to liberation for the individual. Pure Land (*Sukhavati*) Buddhism, attributed to Buddha Amitabha (or Amida in Japanese, who may not be a historical figure), is a derivation from *Mahayana*: it provides a path to enlightenment without the rigours of meditation and other ascetic practices. The 'Pure Land' is the equivalent of the Christian heaven.

For followers of this religious philosophy there are three special objects of veneration, the *Three Jewels of Buddhism*: these are the Buddha, the *Dharma* (Sanskr.) or *Dhamma* (Pali) (his teaching) and the Sangha (the community of practicing Buddhists). Aspiring Buddhists (*bodhisattvas*) must undertake to observe the Five Precepts (*Panca Sila*) – to abstain from harming living beings, from taking things not freely given, from sexual misconduct, from false speech, and from taking intoxicating drinks or drugs. Again, these are precepts that we could all follow in a holistic life. Those who lead the communal life in the Sangha have additional Precepts to observe.

2. The most important book of Kabbalistic literature is the *Sefer Yetsirah* (Book of Creation) that had its origins probably in the early centuries CE [186, p.36]. As the title indicates it deals with problems of cosmogony (the myths or theories of Creation) and cosmology (the structure and functioning of the universe). Another key text is the *Sefer ha-Zohar* (Book of Splendour) compiled from older sources by Moses de Leon (*ca.*1240-1305) in Granada, which, at the time, was the last remaining centre of Judaism and Islam in Spain before it was conquered by the invading Christians. The *Zohar* deals with the mysteries of creation, the role of the *sephiroth*, and the nature of evil. The *Sefer ha-Ikharim* (Book of Fundamental Principles), compiled by Joseph Albo (1380-1444) of Daroca in Spain, is a defence of the greater worthiness of Judaism in the face of the onslaught of Christianity, which was gaining many 'converts' through the threat of execution of non-believers. Other key Kabbalistic scriptures are *The Guide for the Perplexed* by Moses ben Maimon, Maimonides, of Cordoba, who urged believers to seek a higher communion with the Divine than the mere observance of ritual or acquisition of knowledge [Jeremiah 9.23-24]; and *The Fountain of Life* by Solomon ibn Gabirol (1021-1057), written in Arabic as a Platonic dialogue between master and pupil, but discussing the Aristotelian principles of matter and form:

> In seeking Wisdom, the first stage is silence, the second listening, the third remembrance, the fourth practicing, the fifth teaching.
> Solomon ibn Gabirol

Solomon Gabirol also was born and lived in al-Andalus. Theological philosophers from Judaism as well as those from Christianity and Islam living in this medieval period expended most of their efforts in trying to find compatibility between the scriptures that were regarded as having been created from divine revelations and the persuasive rationalisation of the philosophers of ancient Greece and Rome. In taking the title of Maimonides' work, *A Guide for the Perplexed*, as his own, E.F. Schumacher (see Chapter 6) tried to point up the fallacy of a too rationalistic and materialistic attitude to life that paid insufficient regard to matters spiritual.

As explained above, the *Yetsirah* tells of God's Creation of the universe using 32 secret paths of knowledge. These are represented symbolically on a diagram denoted as the *Otz Chiim* (Tree of Life), which can be used as a focus of meditation, rather like the Eastern mandalas. In the diagram there are 10 nodes representing the ten emanations or *sephiroth* from *En Sof Or*, the divine unity. Between these nodes are 22 connections that are denoted by the 22 letters of the Hebrew alphabet; these, together with the 10 nodes, make up the 32 secret paths. The letters were also used to represent numerals, as with the Roman numbering.

The *sephiroth* are represented in various ways. One interpretation is that they are luminous rays of blue light emanating from the divine throne of sapphire (Hebrew: *sappir* – sapphire) [Exodus 24.10; Ezekiel 1.26]. The tenth or *sephirah* is the *Schechinah*, the spirit of God infused into all creation. In some interpretations, the *Schechinah* is viewed as the feminine element of the Divine. Another interpretation of the *sephiroth* is that they can be envisaged as curtains, hiding the Divine Light to create almost darkness. This is the darkness of the material world that some religious dualists, like the Manicheans, regarded as evil. The curtains can be successively lifted in the course of enlightenment by meditation on each *sephirah* in turn. The very word 'enlightenment' means 'letting in the light', a metaphor for 'acquiring wisdom and understanding of the spiritual reality'.

The Kabbalist Rabbi Isaac Luria (1534-1572) envisaged the Divine as Infinite Light:

> 'Know that before the emanations were emanated and the created were created, the exalted and simple Light filled the entire existence, and there was no empty space whatsoever. . . . The universe was created out of nothingness from a single point of light. This nothingness is called the Endless World. The Endless World was filled with infinite Light. The Light was then contracted to a single point, creating primordial space. Beyond this point nothing is known. Therefore, the point is called the beginning. After the contraction, the Endless World issued forth a ray of Light [*Adam Kadmon*]. This ray of Light then expanded rapidly. All matter emanated from that point.'
>
> From the sayings of Isaac Luria, as recorded by his disciple Hayim Vital in his treatise *The Tree of Life* [quoted in [24], pp. 37, 60]

While previous Jewish theologians saw the act of Creation as a positive event, Luria viewed it negatively inasmuch as the Godhead of light had to withdraw within itself (*tzimtzum*) to create the void (*tehiru*) for the universe to occupy – the self-exile of God from infinite space, for which the exile of the Jews from their homeland was symbolic. Creation was then carried out by the divine emanations, the ten *sefiroth*.

The symbol of light in the Jewish religion is the menorah, the seven-branched candlestick fashioned out of gold as described in Exodus [Chap. 25. 31-37]. An eight- or nine-branched version of the candelabrum is used in celebration of the Jewish Festival of Light (*Hanukkah* – dedication), commemorating the rededication of the Temple of Jerusalem. In the Kabbalah, each of the seven branches of the menorah represents one of the *sephiroth*, and the remaining three *sephiroth* are represented by the nodes where the branches meet the central column.

The Tree of Life also features in theosophical and spiritualist beliefs, where the ten *sephiroth* are replaced by archangels – Yesod by Gabriel, Tifereth by Michael, Hod by Raphael, and so on. Spiritual healers associate the seven levels of the Tree of Life with the seven *chakras* (see Chapter 1). There are seven levels in the Tree of Life because six of the ten *sephiroth* occur in complementary pairs. In the colour therapy used by spiritual healers, the blue light of the *sephiroth* is associated with healing. The underlying principle of the stylised representation of the Tree of Life is to focus attention in meditation on the appropriate *sephirah* or archangel to act as a spirit guide in the everyday world: the aim is to lift the curtains of materialist illusion (*maya*) so that we may see spiritual reality. The Kabbalah teaches us that the disquiet of the human spirit originates from desire. This recalls again for us the Buddhist concept of *dukkha* (disquiet, unease) arising from questing after material goals.

3. For those who would need this spiritual concept interpreted in more secular rational terms, there has been a theory amongst cosmologists for several decades now that the material universe we know represents only a few percent of the totality of created matter and that some 95-99% of the universe comprises 'dark matter' and 'dark energy', the nature of which is completely unknown as they are undetectable by any scientific instrumentation yet developed! The 'dark matter' theory is generally attributed to the eccentric Swiss-American astronomer Fritz Zwicky (1898-1974) working in the 1930s [52]. Dark matter exerts a gravitational force but little else. It is in a constant state of dynamic equilibrium with dark energy, that is, the two are constantly inter-converting, like the entities described as subatomic particles. Electromagnetic energy from the sun and stars passes through this medium unaffected. The constitution of dark energy is thought to be the quantum vacuum or zero point field of energy that keeps the planets, stars and galaxies in motion. Some spiritualists believe that this is the domain of the spiritual realm inhabited by the souls of those who have departed earthly life, a domain of the Communal Soul that the mystics have always 'seen'.

4. Until the 20th century, most of what was known about the Gnostics was to be found in the Bible, written by those who regarded Gnosticism as heresy. This situation changed fundamentally after the discovery in December 1945 of a number of manuscripts on papyrus in a cave near the town of Nag Hammadi in Upper Egypt. The manuscripts were Coptic

language translations of still earlier texts that probably originated in the first century after the life of Jesus, at about the same time as the books of the New Testament were written. The Gnostic Gospels give a very different view of Jesus [see also ref. 311].

Coptic Christianity is the Christian movement of Egypt which, to this day, claims some ten million adherents out of a total population of sixty million, most of whom are Muslims. There are other Coptic sects distributed throughout the world. By edict of the Fourth Ecumenical Council of Chalcedon (451 CE), Orthodox Christians must believe that Jesus is both human and divine in one person – the so-called hypostatic union. The Coptic Church found this decree unacceptable and they hold a monophysite belief – Jesus is divine in nature but of human form, and the two natures are united in the one being 'without mingling, without confusion, and without alteration' [from the Declaration of Faith at the end of the Coptic divine liturgy].

Many of the ideas contained in the Nag Hammadi manuscripts conflicted with those of orthodox Christianity, like rejection of the virgin birth and bodily resurrection of Jesus. Even the biblical gospels relate that Jesus appeared to his disciples 'in another form' so that they did not recognize him [Mark 16, 12; Luke 24, 12-16]. Mary Magdalene similarly thought he was the gardener when he appeared to her [John 20, 10-15]. So even orthodox biblical sources are equivocal about the issue of whether the resurrection of Jesus was spiritual or corporeal. This story of Mary Magdalene's encounter with Jesus also makes her the first to experience the resurrected Jesus, explicitly stated also in Mark's gospel [Mark 16. 9]. The Catholic Church affirms Peter's right to be recognised as the first leader of the Church, partly on the authority of Jesus himself [Matthew, 16: 16-19] and partly on the grounds that, according to some authorities, it was he rather than Mary Magdalene who was supposed to have been the first to see the risen Christ [280, pp.39-41]. The Gnostic text of the *Apocalypse of Peter* affirms that Peter was in a deep trance when he saw Jesus who is reported to have said 'I am the intellectual spirit, filled with radiant light' [*Apocalypse of Peter* 83, 8-10]. This does not read like an account of corporeal resurrection. The author of this text also contemptuously dismisses the authority of Church officials: 'Others . . . outside our number . . . call themselves bishops and also deacons as if they had received their authority from God . . . These people are waterless canals' [*Apocalypse of Peter* 79. 22-30; quoted in 280, p.65]. The availability of these manuscripts has inspired works of fiction that, not surprisingly, have incensed some in the orthodox Christian faith, but their popularity demonstrates the ongoing human need for perpetuation of myths.

The Gnostic *Gospel of Thomas* suggests that Jesus had a twin brother, Judas Thomas; even the New Testament suggests that Jesus had several siblings [Mark 6.3] including the twin who is there called Juda or Judas [Matthew 13, 55]. The *Gospel of Philip* implies that Jesus had a much more intimate relationship with Mary Magdalene than is to be found in the orthodox Bible. There is a story that Mary Magdalene left the Holy Lands after the death of Jesus and settled in Southern France: in the medieval village of Vézelay there is a basilica of Sainte Marie-Madeleine that is said to contain the worldly remains of Mary. This of course may very well be another 'Holy Shroud of Turin' myth or hoax!

It is acknowledged that the disciple Thomas travelled to India and founded a Christian Church there. Did Jesus survive the crucifixion and was he among the lost ten tribes of Israel reputed to have been taken to the East by the Assyrians – or did he perhaps follow them? This was suggested by the Russian writer Nicolas Notovitch in his 1894 book *The Unknown Life of Jesus*, but we have no independent evidence as to whether or not this a piece of fiction or another deliberate hoax. It is based on a manuscript that the author said he had found in a Hemis monastery in Leh, the capital of Ladakh in Nepal. This account is supposed to fill in the long gap between the ages of 14 and 29 in the life of Jesus, of which there is surprisingly no mention in the gospels. It also accounts for the relatively short time of nine hours [Matthew 27, 46; Mark 15, 34; Luke

23, 44] for which Jesus was left on the cross, suggesting that Jesus might have been still alive when he was taken down from the cross. When Jesus' tomb was found to be empty it is conjectured that he was resuscitated rather than resurrected and, after he had fully recovered, travelled to Northern India for safety. People are known to have survived crucifixion, if rescued early enough. If this were so, it would make nonsense of the claim by the Christian Church of Jesus sacrificing his life to save humankind from sin. In Kashmir there is a legend of a prophet called St. Issa, also known as Yus Asaf who arrived in Kashmir at around 30 CE. The local scripture, the *Bhavishyat Mahapurana* [vol. 9, verses 17-32], contains an account of events in the life of Issa-Masih (Jesus the Messiah). He is supposed to be buried in a mausoleum underneath a Muslim prophet, Syed Nazir-u-Din. As the tomb is sacred, there is no possibility of scientific examination. The book by Holger Kersten [202], a theologian from a Bible Studies College in Freiburg, is another better researched and much more recent account of the same events, which gives them greater credibility. But whether they are fact or fiction, like other principles of Christianity, remains a mystery.

Another Gnostic gospel, the *Gospel to the Egyptians*, claims to be the sacred book of the Great Invisible Spirit, presenting a much more feasible spiritual view of the Divine than the anthropomorphic imagery of God that has been the norm in the past. Orthodox Judaism, Christianity and Islam all view God as utterly unattainable to the human realm – a chasm separates Man and God. But some of the Gnostic gospels talk of God within Man, and how ultimate self-knowledge (knowing our true spiritual purpose and destiny) is knowledge of God. For these views, and the idea that God was a non-real paragon of all that is good created by the human mind, the cleric Anthony Freeman was deprived of his living in a Sussex parish. In her authoritative text on the Gnostic gospels Elaine Pagels writes: 'when the disciple attains enlightenment, Jesus no longer serves as his spiritual master: the two have become equal – even identical' [280, p.19]. Again, we recall the idea that the bodhisattva can aspire to reach the status of Buddha, alongside Gautama, once he has reached *nirvana*. Whether Buddhist beliefs influenced the writers of the Gnostic gospels, or whether the two streams of thought emerged independently is not clear. Earlier Greek philosophers were almost certainly aware of the ideas of eastern mysticism.

5. For all Muslims, the Five Pillars of the faith are fundamental: *shahadah* – the declaration of faith that there is no God but Allah (*tawhid*) and the Prophet is His messenger; *salah* – prayer, with washing of hands, feet and face beforehand, and observance of the principles of Islam in word and deed; *zakah* – alms giving; *sawm* – fasting in the month of Ramadan; and *hajj* – pilgrimage to Mecca (*Makkah*) at least once in a lifetime; as prescribed in the Koran (*Qur'an*). The practical observance of these ritual duties is called *fiqh* and is obligatory for all Muslims. Sufi *tasawwuf* concentrates on the intentions behind the rituals and how these should affect behaviour in everyday life to let Allah into the heart, mind and soul (*dhikr*). Sufism arose in the 9th and 10th centuries because some followers of the faith were uneasy about such ritualistic practices becoming institutionalised by the state; expression of the faith was held to be a personal experience. Furthermore, they deplored the excessive luxuries and lax morality of the Caliphate in Baghdad and Damascus.

6. Abraham, Moses, Jesus and the prophets of Israel are all considered prophets within Islam but, for Muslims, Muhammad was the last. The early years of Islam after the death of the Prophet in 632 were characterized by bloody power struggles. The leadership of the faith passed successively to Abu Bakr (632-634), Umar (634-644), Uthman (644-656) and thence to Ali. After his murder in 661, Mu'awiyah established the Umayyad dynasty, but the rule of the Umayyads in Syria lasted less than a century. The first dynasty that began to rule after the death of Ali were massacred by the Abbasids in 750. Fortunately, just one of the Umayyad dynasty, Abd al-Rahman, survived to reappear in Spain in 755 CE and the subsequent success of al-Andalus must be credited to him and his Umayyad successors. With the cultural, economic, and academic success

of the Muslim territory of al-Andalus (modern Spain and Portugal), from the 8th to 11th centuries, it was effectively independent of the centre of the faith in Damascus. As part of this cultural development, Jewish, Christian and Islamic scholars worked side by side to make available to the world the scholarship of the ancient Greek rationalists. It is from this historical development that the learning of the ancients was passed down to Christendom in the West through the development in the 12th and 13th centuries of the universities.

7. Medieval Islamic scholarship produced a quartet of great religious philosophers, though only one who was overtly Sufist. The earliest Arabic philosopher was Abu Nasr al-Farabi (870-950), who was born in Turkestan and studied and worked in Baghdad. It was he who was primarily responsible for blending the wisdom of the Greek rationalists with Islamic philosophy, though in this he was preceded by the less well known Abu Yusef al-Kindi (*ca.*800-870), most of whose output is lost. This task of trying to achieve compatibility between their own divinely revealed scripture and the widely-known works of pagan Greek rationalists became a preoccupation of theologians from Judaism, Christianity and Islam during the medieval period. al-Farabi was known throughout Islam as the 'Second Master', the first being Aristotle. He was fortunate in being the first Islamic philosopher to have available to him full Arabic translations of many of the great works from classical antiquity. His main interests lay in logic and political theory.

Ibn Sina or Avicenna (980-1037) was a philosopher, physician and student of the natural sciences. He was born near Bukhara in what is now Uzbekistan. One of his best known works is his *Kitab al-Shifa* (*Book of Healing*), that is, healing for the soul, and a medical textbook, *al-Qanun fi'l-Tibb* (*The Canon of Medicine*), which remained in use until the Renaissance. His late treatise on philosophy *Isharat wa-Tanbihat* (*Demonstrations and Affirmations*) is one of the few of his works that have been preserved in their entirety.

The man who was probably the greatest of the academic mystical philosophers was Abu Hamid al-Ghazzali (1059-1111), a professor of theology and law in Baghdad, though born at Tus in the northern part of Iran. It was largely because of his mystical writings, like *The Niche of the Lights*, that Sufism became an established movement within Islam. The *hujjat al-Islam* (*Proof of Islam*) is one of the most important works of Islamic theology after the Koran. His autobiography describing his adoption of Sufism has been compared to the influence of the *Confessions* of Augustine in Christianity. The principles of Islam as he saw it are contained in his work, *The Revival of the Religious Sciences*. His *Incoherence of the Philosophers* was an attack on the philosophical principles of Aristotle and Plato, whose works were now becoming much more widely known at this time. This book, with its criticism of rationalism as a path to religious belief, gradually led to a withdrawal of Islam from the contribution it had made over the preceding four centuries to knowledge in science and philosophy. However, he maintained that mystical experiences must be translated into good works for others on the mundane plane of existence, and the devotion of the humble to ethical behaviour was as important as the philosophical contributions of the academic elite.

Finally of this quartet there was another great medieval Islamic philosopher, Ibn Rushd or Averroes (1126-1198). al-Ghazzali's mystical approach was countered by Averroes, who came from Andalusian Cordoba. In the days when it was possible to be a polymath and make significant contributions to a number of subjects, Averroes was a philosopher, physician and religious judge (*qadi*). He wrote his *Incoherence of the Incoherence* supporting the classical philosophy of the Greek rationalists, which he regarded superior to religion in gaining knowledge of the world, but he saw no incompatibility between philosophy and religion. He was driven out of Spain by the advancing Christian armies and died in exile in Marrakesh in Morocco. The great Jewish philosopher Moses ben Maimon, Maimonides (1135-1204), was one of his pupils. For further details of Islamic philosophers, see refs.

121, 139.

Chapter 4

1. Trees do not seem to use tannins for any purpose other than to repel insect invaders. They cannot therefore afford to expend valuable energy producing these and similar deterrent compounds if they are not needed, so they have a mechanism whereby the repellents are manufactured when required. David Rhoades studied the defences mounted to infestation in red alder and sitka willow in the western United States and found that the trees made amounts of another toxin, proanthocyanidin, in their leaves to deter caterpillars with which the trees had been deliberately infected [312].

2. We know of the widespread distribution of gardens in Roman civilization from the results of the excavation of Pompeii after it was smothered by volcanic ash from the eruption of Vesuvius in 79 CE.

3. An account of the origins and use of the ogham alphabet can be found in an essay *On the Ogam Beithluisnin* by the mid-19th century Bishop of Limerick, Charles Graves.

4. The *I Ching* comprises a book of 64 hexagrams, figures made out of a column of six lines, either broken or unbroken. The broken lines – – represent the feminine *yin*, while the unbroken lines — represent the masculine element *yang*. Each hexagram is a combination of two trigrams, and each of these represents a natural element: *ch'ien* – heaven (for creative power and enduring strength), *k'un* – earth (for receptive power), *k'an* – water (for skill, power to overcome obstacles), *li* – fire (for warmth and wisdom), *chen* – thunder (for energy and inspirational power), *ken* – mountain (for completion and new beginnings), *sun* – wind (for adaptive power) and *tui* – lake (for enthusiasm and persuasive power). The sage would throw three coins, whose faces represent *yin* and *yang*, six times and from each of the *yin/yang* combinations would construct one row of the hexagram; the resulting pair of trigram combinations would then be used to predict future events or personal character-

istics in the life of the subject.

5. The large blocks and their lintels weighing up to 50 tonnes each at Stonehenge and Avebury are made of a variety of sandstone known as sarsen. This rock occurs naturally in the Marlborough Downs some 30 kilometres away. The smaller stones of the circles are called bluestones from their blue-gray hue when wet and they must have been transported nearly 400 kilometres from the mountains of Presceli in South Wales, for no such rocks as this outcrop anywhere closer. The only other site for quarrying bluestone would have been in Southern Ireland. Both locations would probably have involved transportation of the stones by sea for part of their journey. The ditch surrounding the stone structure at Stonehenge, with its bank of excavated chalk rubble that is thought to have been some 2 meters high originally, forms a circle of nearly 90 meters in diameter. A line of solitary standing stones, called menhirs, form the West Kennet Avenue at Avebury. This line is more than 2 kilometres long stretching away from the Avebury circle, though most of the stones have disappeared over the past two centuries. The outer circle at Avebury is over 300 metres in diameter and the village of Avebury nestles within it. Within this outer perimeter are two smaller circles, each about 100 metres wide.

Equally impressive as the Avebury circle, but in a different way, and only a short distance away from it is the mound of Silbury Hill, also dating from around 3000 BCE. This huge construction, nearly 40 meters high and 30 meters across its flat top, is now covered with earth and grass, but is made of a series of chalk steps. Of the same date as Avebury, Silbury Hill corresponds to the Neolithic and Early Bronze Ages in Britain. It covers a site of 2¼ hectares. Beside the Avebury circles the remains of an adult and two adolescents have been found, suggesting use of the site also as a burial ground or sacrificial temple. Once again, the importance of religion in the lives of these early peoples is clearly demonstrated by the effort that must have been required to build monuments of this impressive stature.

6. The message of the stones is in accord with what has been said in

a secular context about capitalist economic philosophy by numerous authors in the past few post-war decades (e.g. Galbraith [113], Jenkins [191], Packard [279], Schumacher [324], Rifkin [316]). The origins of pollution in our western society, of which we are all too aware, are cited as the cause of this imbalance – excessive traffic on land, sea and air, much of it carrying food and other supplies back and forth across the globe [273]; the over-zealous use of man-made chemicals as pesticides, fertilizers, growth-promoters [155, 156], colourants, preservatives and flavourants of food [233, 234], to produce ever more food of poorer quality to feed an already over-populated world; constant bombardment by the advertising industry encouraging us to ever greater material excesses [285]; and increasing sources of electromagnetic radiation everywhere [46; 274].

Chapter 5

1. These are represented today by the thirteen red and white stripes of the American flag.

2. It was not until 1833 that William Wilberforce succeeded in ending the slave trade within the British Empire.

3. The position of the English political writer and orator Edmund Burke (1729-1797) was ambivalent concerning the rights of the common man to freedom from aristocratic rule. On the one hand he supported the claims of the American colonists for freedom from rule by the king of England and advocated the establishment of an American republic to replace the rule of British monarchy. On the other hand, his 1790 pamphlet *Reflections on the Revolution in France* supported retention of the hereditary monarchy in place of rule by the people in France. It was this publication especially that provoked Paine to produce *Rights of Man*, criticizing Burke specifically and supporting and praising the French Revolution.

4. In the early part of the 18th century there were said to be over 600,000 slaves held by ministers of the Protestant churches alone in the

southern states of America.

5. Paley expounded his views of religious motivation in his work *The Principles of Moral and Political Philosophy* (1785), in which the famous 'pocket-watch on the heath' argument for the existence of God as Designer of the universe appears. The complexity of a pocket watch implies the necessity for a designer and craftsman for such an intricate mechanism. How much greater then is that need when one considers the complexity of the universe.

6. The concept originated with Clare Graves, but the term was introduced by Don Edward Beck and Christopher Cowan in their book *Spiral Dynamics* [21]. The original motivation for these ideas in America was, not surprisingly, mastery of business and maximization of profit, but the ideas do have application elsewhere, as the title of one of Ken Wilber's publications makes clear: *A Theory of Everything: An integral vision for business, politics, science, and spirituality* [390]. Wilber, an American philosopher and psychologist, has made much of this idea in his book which, as the title indicates, is an essentially practical approach to a holistic world-view.

7. Alexis Charles Henri de Tocqueville (1805-1859) was a French magistrate and percipient commentator on the structure of societies. His masterpieces are his study of American society after independence but before the 1861-65 Civil War, *De la démocratie en Amérique* (1835, 1840), and his study of the dangers as well as moral justice of popular government and social equality in his *L'Ancien Régime et la Révolution.* Although he was a passionate believer in liberty, he did not think that this was compatible with equality.

8. Mark D. Griffiths is a Professor in the Division of Psychology at Nottingham Trent University in Britain. He has over 100 research publications to his name concerning various kinds of addiction and he has warned the government of the dangers to young people of encouraging gambling as a national pastime. Unfortunately, when there is the prospect of extra taxes to fill the Exchequer, long-term dangers to society are a

secondary consideration for government [see, for example, refs.137, 138].

Chapter 6

Economists apparently differentiate between GDP and GNP, but for our purposes they can be regarded as identical.

Chapter 7

1. These are generally called cyanobacteria today as less than half of this group are actually blue-green in colour.

2. The group of bacteria called rhizobia lives in and on the roots of legumes like soybean, clover and alfalfa. Fungi-like organisms called actinomyces live symbiotically with nodulated non-legumes, like the alder (*Alnus* sp.), tea tree (*Ceanothus* sp.) and sumac (*Coriaria* sp.). Fungal–algal associations called lichens also play their part in N 'fixation' – converting inert atmospheric N into soluble compounds of ammonium or nitrate that plants can use. Those same organisms then go on to convert these compounds into amino acids, while some other 'denitrifying' bacteria turn the compounds back into atmospheric N to complete the cycle. Nitrogen is also excreted by animals as urea, porphyrins and other N compounds.

3. The physician Galen (*ca.*130-*ca.*200 CE) was the first of note to use plants to treat human illnesses. In the 20th century, the English medical practitioner, Edward Bach (1880-1936), developed a method of homeopathic treatment of maladies by the use of extracts of flowers. He published details of his method in his book *The Twelve Healers* [10]. Bach was born in Moseley, near Birmingham in England and studied conventional medicine in London before becoming a homeopathic healer. As with other homeopathic treatments there is little or no measurable content of active ingredient in the Bach remedies and the mechanism of these and other homeopathic remedies is highly controversial.

Chapter 8

1. The commercial use of the numeral '4' instead of 'for' and the numeral '2' instead of 'to' or 'too', the abbreviation 'R' for 'are' are some specific examples that encourage this slovenly use of the language. In addition, it is confusing for very young children who are in the process of learning how to use the language correctly. Words in any language evolve but these abbreviations are not even words. Repeatedly we see the apostrophe with the letter 's' misused on television, and often hear grammatical errors like 'There's many examples . . .' It's heartening to see how popular Lynne Truss's book, '*Eats Shoots and Leaves*' [372], became in the year or two after its publication showing that there are still many people who care about the proper use of our language.

2. The ratio of the frequencies of vibration of notes that form a musical octave is 2:1; that of the perfect fifth is 3:2 and that of the perfect fourth 4:3. These were the only harmonies permitted in medieval church music.

3. An 8th century Arabic scholar, al-Khwarizmi, wrote a treatise called *Al-jabr wa'l muqabalah,* which translates approximately as 'The Rules of Restoration and Reduction', that became so popular with scholars throughout Europe that its shortened title became the name of its subject matter. Of course, this was not the start of algebra, for some such system was known to the Babylonians, Greeks, Hindus and Chinese.

4. Although he had been born in Pisa, Leonardo Fibonacci was brought up in Algeria where his father was a customs inspector: hence his need to learn Arabic. Fibonacci was known throughout the Islamic and Christian world because of his book *Liber abaci* (*Book of the Abacus*). Fibonacci's name is known to schoolchildren today from the sequence of numbers that bears his name. The sequence is formed by adding the two previous numerals together to get the next (1, 1, 2, 3, 5, 8, 13, 21 . . .). This series has some remarkable properties. When we divide each number by the preceding one, we get closer and closer to 1.618, described as the 'limiting ratio' of successive pairs of numbers of the Fibonacci sequence,

sometimes denoted by the Greek letter 'phi', ϕ. We have represented it by 1.618 but actually it is an irrational number (one whose digits go on forever) and, for the mathematically inclined, it is *exactly* $\frac{1}{2}(1+\sqrt{5})$. Readers might like to work this out on their calculators. ϕ is the only number for which the relationship $\phi = 1 + 1/\phi$ is true.

The dimensions of the Great Pyramid at Gizeh, and others, were constructed to conform to this number, one of the 'mystical numbers' of mathematics in the ancient world, the so called Golden Section or 'sacred ratio'. This ratio was subsequently used in many great architectural works of ancient Greece and, in more recent times, by designers such as Leonardo da Vinci and Christopher Wren. The Papyrus of Ahmes from ancient Egypt, which is now housed in the British Museum in London, describes the building of the Great Pyramid at Gizeh in 4700 BCE. Ahmes's account describes how the sacred ratio was used over and over again in calculating dimensions within this structure.

The occurrence of the Fibonacci series is not limited to man-made structures. It is seen in the botanical property of phyllotaxis – the arrangement of leaves on the stem of a plant. In ways too complicated to go into here in any detail, Fibonacci ratios are also to be found in parastichies (helical lines) that can be drawn in the arrangement of scales on a pine cone or a pineapple, or the seeds in a sunflower head. Here again we have another example of nature holism – a connection between the abstract world of mathematics and the world of Nature. One reason why the regular pentagon featured as a symbol in many pagan rituals is that the ratio of the length of a diagonal to that of a side is again 1.618... The same ratio occurs in the pentagram or five-sided star, another pagan symbol.

5. The fraction one-third expressed as a decimal is 0.333… indefinitely; two-thirds is 0.666… indefinitely. With the elevenths, two digits oscillate, but always add up to 9: thus 1/11 = 0.0909…, 2/11 = 0.1818…, 3/11 = 0.2727…, while 8/11 = 0.7272…, and so on. With the sevenths, we have a block of six repeating numbers: 1/7 = 0.142857142857…, each

block always adding up to 27. It used to be part of the examination syllabus to express numbers from our decimal system in binary format: thus, 8 = 1000, 15 = 1111, and so on. Binary numbers are important in computer programming. All students should have learnt by rote that the cube roots of small numbers like 8 and 27 are 2 or 3. Once they have mastered quadratic equations, they are in a position to find the *other two* cube roots of 8 and 27. There are numerous simple problems like this that the diligent student will find both interesting and stimulating, as indicated by the current interest in the old Japanese game of Sudoku.

REFERENCES AND SOURCE BIBLIOGRAPHY

1. Alcoholism. www.wrongdiagnosis.com.

2. Alternative medicine. Articles in *The Times* newspaper for Tuesday, 23 May 2006.

3. American education statistics. http://www.rand.org/ pubs/ research_briefs/RB8009/index1.html

4. Aquinas, Thomas, *Summa Theologiae, A Concise Translation*, edited by Timothy McDermott; Eyre and Spottiswoode, 1989; Methuen, 1991, London. Originally published 1266-73.

5. Archer, C.I. and M.Z. Jacobson, The spatial and temporal distributions of U.S. winds and wind power at 80m derived from measurements, *J. Geophysical Research*, 16 May 2003.

6. Armstrong, Karen *A History of God*, Heinemann, London, 1993.

7. Armstrong, Karen, *A Short History of Myth*, Canongate, Edinburgh, 2005.

8. Armstrong, Karen, *The Great Transformation: The beginning of our religious traditions*, Alfred A. Knopf, New York, 2006.

9. Astley, Jeff, David Brown and Ann Loades, *Problems in Theology 4. Science and Religion*, T & T Clark, London, 2004.

10. Bach, Edward, *The Twelve Healers and Other Remedies*, C.W. Daniel, Rochford, Essex, UK, 1933.

11. Backster, Cleve, Evidence of a primary perception in plant life, *Intern. J. of Parapsychology*, **10**(4), 329-48, (1968).

12. Bagdikian, Ben, *The Media Monopoly*, 1983; 2nd enlarged edition called *The New Media Monopoly*, 2004, Beacon Press, Boston, Massachussetts.

13. Bailey, Jarrod, Non-human primates in medical research and drug development: a critical review, *Biogenic Amines – Stress and*

Neuroprotection, **19**(4-6), 235-55, Dec. 2005, VSP (Brill).

14. Bailey, Jarrod, A. Knight and J. Balcombe. The future of teratology research is *in vitro*. *Biogenic Amines – Stress and Neuroprotection*, **19**(2), 97-145, 2005, VSP (Brill).

15. Bailey, Ronald, *ECOSCAM: The false prophets of ecological apocalypse*, St. Martins Press, 1993.

16. Bailey, Ronald (ed.), *The True State of the Planet*, The Free Press, New York, 1995.

17. Balcombe, Jonathan, N.D. Barnard and S. Sandusky, Laboratory routines cause animal stress, *Contemporary Topics in Laboratory Animals*, **43**(6), 2004.

18. Barenboim, Daniel, from the first Reith Lecture, quoted in the BBC *Music Magazine* of June 2006, p.13.

19. Barnet, Richard J. and John Cavanaugh, *Dreams: Imperial corporations and the new world order*, Simon and Schuster, New York, 1994.

20. Barrow, John D. and Frank J. Tipler, *The Anthropic Cosmological Principle*, Oxford University Press, Oxford, 1986.

21. Beck, Don Edward and Christopher Cowan, *Spiral Dynamics: Mastering values, leadership, and change*, Blackwell, Cambridge MA, 1995; Oxford, UK, 1996.

22. Behe, Michael J., *Darwin's Black Box: The biochemical challenges to evolution*, The Free Press, New York, 1996.

23. Bentham, Jeremy, *An Introduction to the Principles of Morals and Legislation*, 1789.

24. Berg, Yehuda, *The Power of Kabbalah*, Hodder Mobius, London, 2004.

25. Bergquist, Lars, *Swedenborgs Hemlighet*, Bokförlaget Natur och Kultur, Stockholm, Sweden, 1999; *Swedenborg's Secret*, English trans. by Bergquist for the Swedenborg Society, London, 2005.

26. Berkeley, George, *The Principles of Human Knowledge*, 1710, edited by G.J. Warnock, Collins (Fontana), London, 1962.

27. Berry, Thomas, *The Great Work: Our way into the future*, Bell

Tower, New York (Random House), 1999.

28. BPH (Benign prostate hyperplasia).

(a) See the British Medical Journal website and further references therein at

http://www.besttreatments.co.uk/btuk/conditions/8848.html.

See also Edzard Ernst, The risk-benefit profile of commonly used herbal therapies, *Annals of Internal Medicine*, **136**(1), 1 January 2002 (includes saw palmetto). (b) Syed Bilgramia and Bernard Greeberg, Commentary, *The Lancet*, **344**, 700-1, 1994.

29. Blaukopf, Kurt, *Mahler*, Thames and Hudson, London, 1976.

30. Blythe, Sally Goddard, *The Well-Balanced Child*, Hawthorn Press, 2004.

31. Bohm, David, *Wholeness and the Implicate Order*, Routledge and Kegan Paul (Ark), London, 1983.

32. Bowker, John (editor), *The Oxford Dictionary of World Religions*, Oxford University Press, Oxford, UK, 1997.

33. Braden, Gregg, *The Isaiah Effect*, Hay House, Carlsbad, CA, 2000.

34. Brittan, Samuel, *Capitalism With A Human Face*, Edward Elgar, Aldershot, UK, 1995.

35. Broom, Neil, *How Blind is the Watchmaker*? Intervarsity Press, Downers Grove, IL, USA, 2001.

36. Broughton, Richard, *Parapsychology: The controversial science*, Ballantine Books, USA, 1991; Rider (Random Century), London, 1992.

37. Brown, H. Morrow, *The Asthma and Allergy Reference Book*, Harper & Row, London, 1985.

38. Brown, Lester R., *Outgrowing the Earth*, W.W. Norton, New York, 2004. Lester Brown *et al.* also edit an annual publication of the WorldWatch Institute since 1992 that surveys the health of the Earth in its basic components of air, water and soil.

39. Byrd, R.C., *Southern Medical Journal*, **81**(7), 826-829, 1988.

40. Caldwell, J. Species differences in metabolism and their toxico-

logical significance, *Toxicology Letters* **64**, 106, 1992.

41. Campbell, Joseph, *The Power of Myths*, Doubleday, New York, 1988.

42. Campbell, T.C. and Junshi, C. Diet and chronic degenerative disease perspectives from China, *Amer. J. Clin. Nutrition*, **59** (supplement), 11.535-11.615, 1994.

43. *Cancer Trends in England and Wales* 1950-1999, Office for National Statistics, 2005.

44. Capra, Fritjof, *The Tao Of Physics*, Wildwood House, London, 1975.

45. Capra, Fritjof, *The Hidden Connections*, Harper/Collins, London, 2002.

46. Carson, Rachel, *Silent Spring*, Houghton Mifflin, New York, 1962.

47. Carter, Brandon, Large number coincidences and the anthropic principle in cosmology, in *Confrontation of Cosmological Theories with Observation*, M.S. Longair (ed.), Reidel, The Netherlands, 1974.

48. Chandler, Alfred, *Scale and Scope of Industrial Capitalism*, Harvard University Press, 1990.

49. Chodorow, Nancy, *The Reproduction of Mothering*, University of California Press, Berkeley, California, USA, 1978.

50. Chodorow, Nancy, *Psychoanalytic Theory and Feminism*, Polity Press, Cambridge, UK, 1988.

51. Chopra, Deepak, *The Seven Spiritual Laws of Success*, New World Library and Amber-Allen Publishing, California, 1994.

52. Chown, Marcus, *Afterglow of Creation*, Arrow Books, London, 1993.

53. Cicero, Marcus Tullius, *Offices*.

54. Clarke, Chris, *Living in Connection: Theory and practice of the new world-view*, Creation Spirituality Books, Warminster, UK, 2002.

55. Cleghorn, C.L. *et al. Europ. J. Clinical Nutrition*, **57**, 170-6, 2003.

56. Cohen, Robert, *Milk: The deadly poison*, Bookworld Press U.S., 1998.

57. Colchester, Alan and Nancy Colchester, *Lancet* **366**, 856 (2005); *New Scientist* 10 Sept 2005.

58. Coleman, Daniel, *Emotional Intelligence*, Bantam, New York, 1995.

59. Coleman, James, *Foundations of Social Theory*, Harvard University Press, 1990.

60. Cooke, Deryck, *The Language of Music,* Oxford University Press, Oxford, UK, 1959.

61. Cooke, Grace, *Meditation*, The White Eagle Publishing Trust, Liss, Hampshire, UK, 1955.

62. Cooper, Diana, *A Little Light on the Spiritual Laws*, Hodder and Stoughton, London, 2000.

63. Cortens, Theolyn, *Living with Angels: Bringing angels into your everyday life*, Judy Piatkus, London, 2003.

64. Cox, Caroline, Herbicide Factsheet: Glyphosate (Roundup), *J. Pesticide Reform*, **18**(3), 3-17, 1998.

65. Cox, Caroline, Herbicide Factsheet: Malathion, *J. Pesticide Reform*, **23**(4), 10-15, 2003.

66. Creutzfeldt-Jacob Disease (CJD). *New Scientist*, www.newscientist.com/hottopics/bse/bsetimeline.jsp, 28 December 2001.

67. Crowder, Colin (ed.), *God and Reality: Essays on Christian non-realism*, Mowbray (Cassell), London, 1997.

68. Cupitt, Don, *The Sea of Faith*, BBC Publications, London, 1984.

69. Cupitt, Don, *After All: Religion without alienation*, SCM Press, London, 1994.

70. Darwin, Charles, *The Origin of Species by Means of Natural Selection*, John Murray, London, 1859.

71. Daly, Herman E., *Beyond Growth*, Beacon Press, Boston, Mass., 1997.

72. Daly, Herman E. and Joshua Farley, *Ecological Economics*, Island Press, 2003.

73. Daly, Herman E. and Kenneth N. Townsend (eds), *Valuing the*

Earth, MIT Press, 1993.

74. Davenas, Elizabeth, *et al.*, *Nature*, **333**, 818, 1988.

75. Davies, Paul, *God and the New Physics*, J.M. Dent, London, 1983; Penguin, 1984.

76. Dawkins, Richard, *The Selfish Gene*, Oxford University Press, Oxford, UK, 1976; Paladin (Granada) p.b., London, 1978.

77. Day, Brian, *The Modern Pagan: How to live a natural lifestyle in the 21st century*, Rider (Random House), London, 2006.

78. Dean, Carolyn and Trueman Tuck, *Death by Modern Medicine*, Matrix-Vérité, 2005.

79. De Geus, Arie, *The Living Company*, Harvard Business School Press, 1997.

80. Descartes, René, *Meditations on the First Philosophy*, 1641.

81. Dinwiddy, John, *Bentham*, Oxford University Press, 1969.

82. Donne, John, *Devotions*, 1624, Meditation 17.

83. Dossey, Larry, *Healing Words: The power of prayer and the practice of medicine*, Harper-Collins, 1994.

84. Dossey, Larry, *Reinventing Medicine*, Harper-Collins, 1999.

85. Dossey, Larry, *Healing Beyond the Body*, Shambala Publications, 2001.

86. Durkheim, Emil, *De la division du travail social* (*The Division of Labour in Society*), 1893.

87. Durkheim, Emil, *Le Suicide*, 1897.

88. Dyer, Wayne, *The Power of Intention*, Hay House, Carlsbad, California, 2004.

89. Eccles, John, *How the Self Controls Its Brain*, Springer Verlag, Berlin, 1994.

90. Ehrlich, Paul R. and Anne H. Ehrlich, *Population, Resources, Environment: Essays in human ecology*, W.H. Freeman, San Francisco, 1970.

91. Ehrman, Bart D., *Whose Word Is It?* Continuum (T & T Clark), London, 2006.

92. Environmental Research Foundation, Mad Cow Disease in Humans, *Rachel's Environment & Health Weekly*, No.683, 20 January 2000.

93. Epstein, Isidore, *Judaism*, Penguin, Harmondsworth, Middlesex, UK, 1959.

94. Fagan, Brian M. *People of the Earth*, HarperCollins, London, 1992.

95. Fallon, Sally, *Nourishing Traditions*, New Trends Publishing, Winona Lake, Indiana, revised 2nd edn. 2001.

96. Feinstein, David and Stanley Krippner, *Personal Mythology*, Unwin, London, 1989.

97. Fergusson, David A.S. *The Cosmos and the Creator: An introduction to the theology of Creation.* SPCK, London, 1998.

98. Feuerbach, Ludwig, *The Essence of Christianity*, 1841.

99. Fontana, David, *Is There An Afterlife?* O Books, Ropley, Hants, 2005.

100. Food poisoning. Report on 'Mass Production Promotes Food Poisoning', Burlington Free Press, 10 December, 1997, p.1.

101. Ford, Adam, *Universe: God, Man and Science*, Hodder and Stoughton, London, 1986.

102. Fordham, Frieda, *An Introduction to Jung's Psychology*, Penguin, Harmondsworth, UK, 1953.

103. Fox, Mark, *Religion, Spirituality and the Near-Death Experience*, Routledge, London, 2003.

104. Frankfort, Henri, *Ancient Egyptian Religion*, Columbia University Press, New York, 1948.

105. Freedman, Robert (ed.), *Marx on Economics*, Harcourt Brace 1961, Penguin 1962.

106. Freeman, Anthony, *God In Us*, SCM Press, London, 1993.

107. Freud, Sigmund, *The Interpretation of Dreams*, 1900.

108. Freud, Sigmund, *The Future of an Illusion*, 1928.

109. Freud, Sigmund, *Des Unbehagen in der Kultur*, 1930; translated

by David McLintock, *Civilization and its Discontents*, Penguin, London, 2002.

110. Fromm, Erich, *Escape From Freedom*, Holt, Rinehart and Winston, New York, 1941; published as *The Fear of Freedom*, Routledge & Kegan Paul, London, 1942; Ark (RKP) p.b. 1984.

111. Fromm, Erich, *Man for Himself*, Holt, Rinehart and Winston, New York, 1947; Routledge and Kegan Paul, London, 1949.

112. Fromm, Erich, *To Have or To Be?* Jonathan Cape, London, 1978.

113. Galbraith, John Kenneth, *The Affluent Society*, Hamish Hamilton, London, 1958.

114. Garbarino, James, *Raising Children in a Socially Toxic Environment*, Jossey-Bass (John Wiley), New York, 1995.

115. Gardiner, Alan H. and Kurt Sethe, *Egyptian Letters to the Dead*, London, 1928

116. Gardner, Howard, *Frames of Mind*, Basic Books, New York, 1983.

117. Gardner, Howard, *Multiple Intelligences*, Basic Books, New York, 1993.

118. Garland, C.F. *et al.* Can colon cancer incidence and death rates be reduced with calcium and vitamin D? *Amer. J. Clinical Nutrition* **54**, 192S, 1991.

119. Gelbspan, Ross, *The Heat Is On*, Addison-Wesley, Boston, Mass., 1997.

120. George, Susan and Fabrizio Sabelli, *Faith and Credit: The World Bank's secular empire*, Penguin, Harmondsworth, Middlesex, UK, 1994.

121. Gibb, H.A.R., *Islam*, Oxford University Press, Oxford, UK, 1949.

122. Giddens, Anthony, *Sociology*, Polity Press, Cambridge, UK, 4th edn 2001.

123. Gill, Melanie, The Death of Motherhood, *Daily Mail*, Thursday 20 July 2006.

124. Glaciers. Global Warming? What a load of poppycock, by David

Bellamy, *Daily Mail*, 9 July 2004.

125. Global warming. (a) *Sunday Telegraph* letters section, 23 April 2006; (b) Peter Hitchens, 'Global Warming? It's hot air and hypocrisy', *Mail on Sunday*, 29 July 2001.

126. Goldberg, David J. and John D. Rayner, *The Jewish People: Their history and their religion*, Viking, 1987; Penguin, London, 1989.

127. Goldsmith, Edward, *The Way: An ecological world-view*, Rider (Random House), London, 1992; Themis Books, Dartington, Totnes, Devon, UK, 1996.

128. Goleman, Daniel, *Emotional Intelligence: Why it can matter more than IQ*, Bantam Books, New York, 1995.

129. Gorbachev, Mikhail, *Manifesto for the Earth: Action now for peace, global justice and a sustainable future*, Clairview Books, Forest Row, East Sussex, UK, 2006.

130. Gore, Albert, *Earth In The Balance*, Houghton Mifflin, New York, 1992.

131. Gore, Albert, *An Inconvenient Truth: The planetary emergency of global warming and what we can do about it*, Rodale, USA, 2006 – the book version of the illustrated lecture film.

132. Graham, Phyllis, *The Jesus Hoax*, Leslie Frewin, London, 1974.

133. Gray, John, 'Culture of Containment', *The Guardian*, 20 November 1995.

134. Gray, John, *False Dawn: The delusions of global capitalism*, Granta Books, London, 1998.

135. Gray, S. Breast cancer and hormone-replacement therapy: The Million Women Study. *Lancet*, **362,** 1332, 2003.

136. Griffiths, Bede, *The Marriage of East and West*, Templegate Publishers (US); Collins (UK) 1977.

137. Griffiths, Mark D., Videogame violence and aggression: A review of research, in C. von Feilitzen and U. Carlsson (eds), *Children in the New Media Landscape: Games, pornography, perceptions*, pp.31-33, UNESCO, Sweden, 2000.

138. Griffiths, Mark D., Adolescent gambling: risk factors and implications for prevention, intervention and treatment, in D. Romer (ed.), *Reducing Adolescent Risk: Toward an integrated approach*, Sage, New York, 2004.

139. Guillaume, Alfred, *Islam*, Penguin, Harmondsworth, Middlesex, UK, 1954.

140. Hagen, Steve, *Buddhism, Plain and Simple*, Penguin, Harmondsworth, Middlesex UK, 1997.

141. Hageneder, Fred, *The Living Wisdom of Trees*, Duncan Baird Publishers, London, 2005.

142. Hamblin, Henry Thomas, *The Way of the Practical Mystic*, Polair Publishing, London, 2004.

143. Hampson, Daphne, On being a non-Christian realist, in *God and Reality*, Colin Crowder (ed.), Mowbray (Cassell), London, 1997.

144. Hansard, Christopher, *The Tibetan Art of Serenity: How to heal fear and gain contentment*, Hodder Mobius, London, 2006.

145. Hansen, Helle Rüsz and Spiros A. Pergantis, *J. Anal. Atom. Spectrometry*, **21**, 731, 2006.

146. Hanssen, Maurice, *E for Additives*, Thorsons, 1984; 2nd revised edn. Thorsons (HarperCollins), London, 1988.

147. Happold, F.C. *Mysticism*, Penguin, London, 1963.

148. Hardell, Lennart and Mikael Eriksson, A case-control study of non-Hodgkin lymphoma and exposure to pesticides, *Cancer* **85**(6), 1353-1360, 1999 (March 15).

149. Hardy, Alister, *The Spiritual Nature of Man: A study of contemporary religious experience*, Oxford University Press, Oxford, UK, 1979.

150. Harpur, Tom, *Life After Death*, McClelland and Stewart, Toronto, Ontario, Canada, 1991.

151. Harpur, Tom, *The Pagan Christ: Recovering the lost light*, Thomas Allen, Toronto, 2004.

152. Harris, W.S., M. Gowda, J.W. Kolb, *et al.*, *Archives of Internal Medicine*, **159**, 2273-2278, 1999.

153. Hartmann, Thom, *The Last Hours of Ancient Sunlight: Waking up to personal and global transformation*, Mythical Research Inc, 1998; Hodder and Stoughton, London, 2001.

154. Hartshorne, Charles, *The Divine Relativity: A social conception of God*, Yale University Press, New Haven, CT, USA, 1948.

155. Harvey, Graham, *The Killing of the Countryside*, Jonathan Cape. London, 1996.

156. Harvey, Graham, *We Want Real Food: Why our food is deficient in minerals and nutrients, and what we can do about it*, Constable, London, 2006.

157. Harvey, Graham and Charlotte Hardman, *Pagan Pathways: A guide to the ancient earth traditions*, Thorsons (HarperCollins), London, 2000.

158. Hawken, Paul, *The Next Economy*, Ballantine, New York, 1983.

159. Hawken, Paul, *The Ecology of Commerce*, HarperCollins, New York, 1993.

160. Hawken, Paul, with Amory Lovins, *Natural Capitalism: Creating the next industrial revolution,* Little Brown, Boston, Mass., USA, 1999.

161. Hayek, F.A., *The Fatal Conceit: The errors of socialism*, Routledge, London, 1988.

162. Headworth, Howard, *The al-Andalus Chronicle*, Athena Press, London, 2004.

163. Healthcare in the USA. *Journal of the American Medical Association*, **284**(4), 483-5, 2000, 26 July, article by Barbara Starfield.

164. Helvarg, David, *The War Against The Greens*, Sierra Club Books, San Francisco, 1994.

165. Helvarg, David, *Blue Frontier: Saving America's living seas*, W.H. Freeman, New York, 2001.

166. Heritage, J. The fate of transgenes in the human gut, *Nature Biotechnology*, **22**(2), 170, 2004.

167. Hick, John, *The Problems of Religious Pluralism*, Macmillan,

London, 1985.

168. Hicks, Jerry and Esther Hicks, *The Amazing Power of Deliberate Intent*, Hay House, London, 2006.

169. Hindmarsh, W. Russell, *Science and Faith*, Epworth Press, 1968.

170. His Holiness the XIV Dalai Lama, *The Four Noble Truths*, HarperCollins (Thorsons), 1997.

171. His Holiness the XIV Dalai Lama, *The Universe in a Single Atom: How science and spirituality can serve our world*, Doubleday, New York, 2005; Little, Brown, London, 2006.

172. Hobbes, Thomas, *Leviathan*, 1651; edited by John Plamenatz, Fontana (Collins), London, 1962.

173. Hobson, Peter, *The Cradle of Thought: Exploring the origins of thinking*, Macmillan, London, 2002.

174. Hoffman, Natasha, with Hamilton Hill, *Let the Standing Stones Speak*, O Books, Winchester, UK, 2004.

175. Honoré, Carl, *In Praise of Slow: How a worldwide movement is challenging the Cult of Speed*, Orion Books, London, 2004.

176. Horney, Karen, *Our Inner Conflicts*, Norton, New York, 1945.

177. Hospital infections. Programme on hospital infections 23 Sept 2004, BBC1.

178. Howard, Vyvyan, Synergistic effects of chemical mixtures: Can we rely on traditional toxicology? *The Ecologist*, **27**(5), 192-195, 1997.

179. Humphreys, Christmas, *The Buddhist Way of Life*, George Allen and Unwin, London, 1969.

180. Humphreys, Christmas, *The Wisdom of Buddhism*, Rider, London, 1970.

181. Humphreys, Christmas, *Buddhism*, Penguin, London, 1990.

182. Hutton, Will, *The State We're In*, Jonathan Cape, London, 1995.

183. Illich, Ivan, *Limits to Medicine*, Marion Boyars, London, 1976.

184. Inglis, Brian, *The Diseases of Civilization: Why we need a new approach to medical treatment*, Hodder and Stoughton, London, 1981.

185. Inglis, Brian, *Coincidence: A matter of chance or synchronicity?*

Hutchinson, London, 1987.

186. Isaacson, Sara, *Principles of Jewish Spirituality*, Harper/Collins (Thorsons), London, 1999.

187. Isserlis, Steven, in an interview with Oliver Condy published in the BBC *Music Magazine* for June 2006, p.23.

188. James, William, *The Varieties of Religious Experience*, Longmans 1902, Penguin 1982.

189. James, William, *Pragmatism*, Longmans, Green and Co., New York, 1928.

190. Jaspers, Karl, *The Origin and Goal of History*, trans. Michael Bullock, Yale University Press, New Haven, CT, 1953.

191. Jenkins, David, *Market Whys and Human Wherefores: Thinking again about markets, politics and people*, Continuum (T & T Clark), London, 2000.

192. Jones, Ernest, *The Life and Work of Sigmund Freud*, 3 vols, Basic Books, New York; Hogarth Press, London, 1953.

193. Jones, Howard A., *The Thoughtful Guide to God*, O Books, Ropley, Hants UK, 2006.

194. Jung, Carl W. *Instinct and the Unconscious*, 1919; *The Collected Works of C.J. Jung,* Routledge and Kegan Paul, London, Vol. 8, *The Structure and Dynamics of the Psyche*, 1960.

195. Jung, Carl W. *Psychology and Religion*, 1938/1940; *The Collected Works of C.J. Jung,* Routledge and Kegan Paul, London, Vol. 11, *Psychology and Religion: West and East,* 1958.

196. Kanner, A.D., Theodore Roszak and M.E. Gomes, *Ecopsychology: Restoring the Earth, healing the mind*, Sierra Club Books, San Francisco, 1995.

197. Kant, Immanuel, *Lectures on Ethics*, trans. L. Infield, London, 1930.

198. Kant, Immanuel, *Critique of Practical Reason*, 1788; trans. S. Körner, Penguin, Harmondsworth, Middlesex, UK, 1955.

199. Kassin, Saul, *Psychology*, Houghton Mifflin, Boston, 1995.

200. Kindred, Glennie, *Sacred Celebrations*, Gothic Image Publications, Glastonbury, Somerset, UK, 2001.

201. Keown, Damien, *Buddhism: A very short introduction*, Oxford University Press, Oxford UK, 1996.

202. Kersten, Holger, *The Unknown Life of Jesus*, 1983 (in German); Element Books, Longmead, UK, 1986 (Eng. trans.).

203. Klauck, Hans-Josef, *Apocryphal Gospels*, T and T Clark, London, 2003.

204. Koestler, Arthur, *The Ghost In The Machine,* Hutchinson, 1967; Penguin 1989.

205. Korten, David, *When Corporations Rule The World*, Kumarian Press, San Francisco, 1995; 2nd edn. 2001.

206. Korten, David, *The Post-Corporate World: Life after capitalism*, Kumarian Press, San Francisco, 1998.

207. Kosko, Bart, *Fuzzy Thinking: The new science of fuzzy logic*, HarperCollins, London, 1994.

208. Kuhn, Thomas S. *The Structure of Scientific Revolutions*, University of Chicago Press, 1962; 2nd enlarged edn. 1970.

209. Küng, Hans, *Tracing the Way: Spiritual dimensions of the world religions*, Continuum, London, 2002.

210. Coenzyme Q10 (ubiquinone) and statins (a) Lamperti, C. *et al.* Muscle co-enzyme Q10 level in statin-related myopathy, *Archives of Neurology*, **62**(11), 1709-12, 2005. (b) Levy, H.B. and H.K. Kohlhaas, Considerations for supplementing with coenzyme Q10 during statin therapy, *Annals of Pharmacotherapy*, **40**(2), 290-4, 2006.

211. (a) Lao Tzu, *Tao Te Ching*, original translation into German by Richard Wilhelm; translated into English by H.G. Ostwald, Arkana (Routledge & Kegan Paul); Penguin, London, 1985. (b) Lao Tzu, *Tao Te Ching*, translated by John C.H. Wu, Shambhala, Boston, MA, USA, 1989. (c) Lao Tzu, *Tao Te Ching*, translated by Ralph Alan Dale, Watkins, London, 2002.

(The quotations from the *Tao Te Ching* in this book are my own interpretations of the translations by Dale and Wu)

212. Laszlo, Ervin, *The Creative Cosmos: A unified science of matter, life and mind*, Floris, Edinburgh, 1993, p.129:

213. Laszlo, Ervin, *Science and the Akashic Field*, Inner Traditions Publishers, Rochester, Vermont, 2004.

214. Laszlo, Ervin, *The Chaos Point: The world at the crossroads*, Piatkus, London, 2006.

215. Law, M, Plant sterol and stanol margarines and health, *Brit. Med. J.* **320**, 861-4, 2000.

216. Layard, Richard, *Happiness: Lessons from a new science*, Allen Lane, London, 2005.

217. Leakey, Richard and Roger Lewin, *Origins Reconsidered: In search of what makes us human*, Little, Brown and Co., London, 1992.

218. Leakey, Richard and Roger Lewin, *The Sixth Extinction: Patterns of life and the future of humankind*, Doubleday, New York, 1996.

219. Le Bars, P.L. *et al.* A placebo-controlled, double-blind, randomized trial of an extract of *Ginkgo biloba* for dementia, *J. Amer. Med. Assocn.*, **278**(16), 1327-32, Oct 1997.

220. Linn, Susan, *Consuming Kids: The hostile takeover of childhood*, The New Press, New York, 2004.

221. Lipton, Bruce, *Developmental Biology*, **60**, 26-47, 1977; **61**, 153-165, 1977.

222. Lipton, Bruce, *The Biology of Belief*, Elite Books, Santa Rosa, California; Cygnus Books, Llandeilo, Wales, UK, 2005.

223. Listeria website. www.amm.co.uk/newammfiles/factsabout/fa-list.htm

224. Locke, John, *Two Treatises of Civil Government*, edited by Peter Laslett, 2nd edn., Cambridge University Press, 1967.

225. Locke, John, *On Liberty*, London, 1859.

226. Locke, John, *An Essay Concerning Human Understanding*, 1690, edited by A.D. Woozley, Collins, London, 1964.

227. Lorimer, David (Ed.), *The Spirit of Science: From experiment to*

experience, Floris Books, Edinburgh, 1998.

228. Louv, Richard, *Last Child In The Woods: Saving our kids from nature-deficit disorder*, Algonquin Books, Chapel Hill, North Carolina, USA, 2005.

229. Lovelock, James, *Gaia: A new look at life on Earth*, Oxford University Press, Oxford, UK, 1979.

230. Lovelock, James, *The Ages of Gaia: A biography of our living Earth*, Oxford University Press, Oxford, UK; W.W. Norton, New York, USA, 1988.

231. Lovelock, James, The Environment Now and the Gaian Perspective, in *The Spirit of Science*, David Lorimer (ed.), p.129, Floris Books, Edinburgh, 1998.

232. Lucas, Bill and Alastair Smith, *Help Your Child To Succeed: The essential guide for parents*, Network Educational Press, London, 2002.

233. Mackarness, Richard, *Not All In The Mind*, Pan, London, 1976.

234. Mackarness, Richard, *Chemical Victims*, Pan, London, 1980.

235. Mackay, Alan L., *The Harvest of a Quiet Eye: A selection of scientific quotations*, The Institute of Physics, Bristol, UK, 1977.

236. Mander, Jerry, *In the Absence of the Sacred*, Sierra Club Books, San Francisco, 1992.

237. Manning, Richard, *Grassland: The history, biology, politics and promise of the American prarie*, Penguin Books, USA, 1995, p.127.

238. Maramay, Marc and Val Young, *Earth Will Be Reborn*, O Books, Ropley, Hampshire, UK, 2006.

239. Martial (Marcus Valerius Martialis), *Epigrams*

240. Martin, Paul, *Making Happy People: The nature of happiness and its origins in childhood*, Harper/Collins, London, 2006.

241. Marx, Karl, *Critique of the Gotha Programme* 1875.

242. Maypole. Story in the *Independent* newspaper for 1 May 1989.

243. Mayr, Ernst W., *Evolution and the Diversity of Life*, Harvard University Press, Cambridge, Massachusetts, 1976.

244. McCully, Kilmer S. *The Homocysteine Revolution*, Keats

Publishing, New Canaan, Connecticut, USA, 1997.

245. McIntosh, Alastair, *Soil and Soul: People versus corporate power*, Aurum Press, London, 2001; p.b. 2004 (the edition referred to herein).

246. McKibben, Bill, *The End of Nature*, Anchor Books (Random House), New York, 1989.

247. McNeal, James U. *Kids as Customers: A handbook of marketing to children*, Lexington Books, Lanham, Maryland, USA, 1992.

248. McTaggart, Lynne, *What Doctors Don't Tell You: The truth about the dangers of modern medicine*, Thorsons (HarperCollins), London, 1996.

249. McTaggart, Lynne, *The Field*, Element (HarperCollins), London, 2001.

250. Meadows, Donella, Dennis Meadows, Jorgen Randers and William Behrens, *The Limits to Growth*, Universe Books, New York, 1972.

251. Medication errors. www.amcp.org; (b) www.fda.org

252. Menocal, Maria Rosa, *Ornament of the World: How Muslims, Jews and Christians created a culture of tolerance in medieval Spain*, Little, Brown and Company, Boston, MA, USA, 2002.

253. Midgley, Mary, *Evolution as a Religion*, Methuen, London, 1985.

254. Milk Development Council, Dairy Supply Chain Margins 2003-4, August 2004. 255. Milk Production. www.aae.wisc.edu/future /trends_production.htm.

256. Milk Toxicity. (a) www.preventcancer.com. (b) Samuel S Epstein: Potential public health hazards of biosynthetic milk hormones, *Intern. J. Health Services*, **20**(1), 73-84 (1990).

257. Mill, John Stuart, *On Liberty*, London, 1859.

258. Mill, John Stuart, *Utilitarianism*, London. 1861.

259. Millman, Dan, *The Laws of Spirit*, H.J. Kramer & New World Library, California, 1995.

260. Mills, C. Wright, *The Sociological Imagination*, Penguin,

Harmondsworth, Middlesex, UK, 1970.

261. Mohen, Jean-Pierre, *Standing Stones*, Thames and Hudson, London, 1999.

262. Molyneaux, Brian Leigh and Piers Vitebsky, *Sacred Earth, Sacred Stones*, Duncan Baird Publishers, London, 2001.

263. Monbiot, George, *Heat: How to stop the planet burning*, Allen Lane (Penguin), London, 2006.

264. Mumby, Keith, *Allergies: What everyone should know*, Unwin, London, 1986.

265. Murdock, George, *Social Structure*, Macmillan, New York, 1949.

266. Music and memory. *The Times*, Wednesday 20 September 2006; *Daily Telegraph*, Wednesday 20 September 2006.

267. Neale, Jonathan, *What's Wrong With America? How the rich and powerful have changed America and now want to change the world*, Satin Publications (Vision), London, 2004.

268. Netherwood, T. *et al.*, Assessing the survival of transgenic plant DNA in the human gastrointestinal tract, *Nature Biotechnology*, **22**(2), 204, 2004.

269. Neuberger, Julia, *The Moral State We're In: A manifesto for a 21st century society*, HarperCollins, London, 2005.

270. Nietzsche, Friedrich, *Wille zur Macht*, 1901, 1906; English edn. *The Will to Power*, trans. Walter Kaufmann and R.J. Hollingdale, 1968.

271. Noakes, M. *et al. Amer. J. Clinical Nutrition*, **75**, 79-86, 2002.

272. Norberg-Hodge, Helena, *Ancient Futures: Learning from Ladakh*, Random House (Rider), London, 1992.

273. Norberg-Hodge, Helena, Todd Merrifield and Steven Gorelick, *Bringing the Food Economy Home: Local alternatives to global agribusiness,* Zed Books, London, 2002.

274. Nordström, Gunni, *The Invisible Disease: The dangers of environmental illnesses caused by electromagnetic fields and chemical emissions*, O Books, Winchester, Hampshire, UK, 2004.

275. Null, G. *et al. Death by Medicine*, Nutrition Institute of America,

New York, 2003.

276. Ormerod, Paul, *The Death of Economics*, Faber and Faber, London, 1994,

277. Orr, H. Allen, Devolution: why intelligent design isn't, The New Yorker on line, issue of 30 May 2005.

278. Outwater, J.L. *et al.*, Dairy products and breast cancer: the IGF-1 estrogen and BGH hypothesis. *Medical Hypotheses*, **48**, 453-461, 1997.

279. Packard, Vance, *The Waste Makers*, Simon and Schuster, New York, 1960.

280. Pagels, Elaine, *The Gnostic Gospels*, Random House, New York, 1979; Weidenfeld and Nicholson, London 1980; Penguin, London, 1990 (this is the edition referred to in the text).

281. Paine, Thomas, *Common Sense*, 1776.

282. Paine, Thomas, *The Rights of Man*, Part I, 1791; Part II, 1792.

283. Paine, Thomas, *The Age of Reason*, 1794.

284. Paley, William, *The Principles of Moral and Political Philosophy*, 1785.

285. Palmer, Sue, *Toxic Childhood: How the modern world is damaging our children and what we can do about it*, Orion Books, London, 2006.

286. Pascal, Blaise, *Pensées*, published posthumously in 1670.

287. Peacocke, A.R., *Creation and the World of Science*, Oxford University Press, Oxford, UK, 1979.

288. Pesticide Residues Committee, *Annual Report 2003*. http://www.pesticides.gov.uk.

289. Phillips, Tim, A Single Skill Is Not Enough, *Guardian* newspaper, 28 October 1999.

290. Pimental, David, *et al.* Environmental, energetic and economic comparisons of organic and conventional farming, *Bioscience*, **55**(7), 573-582, 2005 (July).

291. Plant, Jane, *Your Life in Your Hands*, Virgin Publishing, London, 2003.

292. Plat, J and R.P. Mensink, Plant stanol and sterol esters in the control of blood cholesterol levels: mechanism and safety aspects, *Amer. J. Cardiology*, **96**(1A), 15D-22D, 2005.

293. Playfair, Guy Lyon, *Twin Telepathy*, Chrysalis Books (Vega), London, 2002.

294. Plimmer, Martin and Brian King, *Beyond Coincidence*, Icon Books, Cambridge, 2004.

295. Polkinghorne, John, *Science and Creation*, SPCK, London, 1988.

296. Polkinghorne, John, *Science and Providence*, SPCK, London, 1989.

297. Polkinghorne, John, *Reason and Reality*, SPCK, London, 1991.

298. Ponting, Clive, *A Green History of the World: The environment and the collapse of great civilizations*, Penguin, Harmondsworth, UK, 1992.

299. Popper, Karl, *Logik der Forschung*, Vienna, Autumn 1934 (imprint, 1935); *The Logic of Scientific Discovery*, Hutchinson, London, 1959.

300. Postman, Neil, *The Disappearance of Childhood*, Vintage Press (Random House), 1994.

301. Poynder, Michael, *Lost Science of the Stone Age*, Green Magic, Sutton Mallet, Somerset, UK, 2004.

302. Pradervand, Pierre, *The Gentle Art of Blessing*, Cygnus Books, Llandeilo, UK, 2005.

303. Pretty, Jules, The real cost of modern farming, *Resurgence*, March/April 2001.

304. Pretty, Jules and Rachel Hine, Reducing Food Poverty with Sustainable Agriculture: A Summary of New Evidence. Final report from the 'SAFE-World' Research Project, University of Essex, Colchester, UK, February 2001.

305. Price, Weston A. *Nutrition and Physical Degeneration*, 1939; 6th edn. Price-Pottenger Nutrition Foundation, La Mesa, California, 2004.

306. Prior, Ronald, *et al., J. Agric. Food Chemistry*, **46**, 2686-2693,

1998.

307. Randolph, Theron G. and Ralph W. Moss, *Allergies: Your hidden enemy*, Thorsons, UK, 1984.

308. Redfield, James, *The Celestine Vision*, Warner Books, New York, 1997.

309. Rees, Kenneth, The tangled skein: the role of myth in paganism, in *Pagan Pathways: A guide to the ancient earth traditions*, Graham Harvey and Charlotte Hardman (editors), Thorsons (HarperCollins), 2000.

310. Reginster, Jean-Yves, *et al.* Long-term effects of glucosamine sulphate on osteoarthritis progression: a randomized placebo-controlled clinical trial. *The Lancet*, **357**, 251-6, 2001.

311. Reynolds, Alfred, *Jesus versus Christianity*, Cambridge International Publishers, London, 1988.

312. Rhoades, David F. Responses of alder and willow to attack by tent caterpillars and webworks, in *Plant Resistance to Insects*, P.A. Hedin (ed.), American Chemical Society Symposium Series, Vol 208, 1983.

313. Richardson, Alex, *They Are What You Feed Them*, Thorsons (HarperCollins), London, 2005.

314. Richardson, W. Mark and Gordy Slack (eds), *Faith in Science: Scientists search for truth*, Routledge, London, 2001.

315. Ridley, Matt, *Nature via Nurture: Genes, experience and what makes us human*, Fourth Estate (Harper/Collins), London, 2003.

316. Rifkin, Jeremy, *Entropy: A new world-view*, Granada (Paladin), London, 1985.

317. Robins, Don, *Circles of Silence*, Souvenir Press, London, 1985.

318. Robins, Don, *The Secret Language of Stone*, Rider, London, 1988.

319. Robinson, John A.T., *Honest to God*, SCM Press, London, 1963.

320. Roszak, Theodore, *The Voice of the Earth*, Touchstone Books, 1992.

321. Russell, Peter, *The Awakening Earth: The global brain,*

Routledge & Kegan Paul, London, 1982; Ark, 1985.

322. Ryan, Frank, *Darwin's Blind Spot: Evolution beyond natural selection*, Houghton Mifflin, New York, 2002.

323. Saddhatissa, H. *The Buddha's Way*, George Allen and Unwin, London, 1971.

324. Schumacher, E.F. *Small Is Beautiful: A study of economics as if people mattered*, Blond and Briggs, London, 1973; Sphere (Abacus) edition, 1974.

325. Schumacher, E.F. *A Guide for the Perplexed*, Jonathan Cape, London, 1977; Sphere (Abacus) 1978.

326. Schwartz, Gary E. with William L. Simon, *The Afterlife Experiments: Breakthrough scientific evidence of life after death*, Atria Books, New York, 2002.

327. Selbourne, David, *The Principle of Duty*, Sinclair-Stevenson, London, 1994.

328. Sen, K.M. *Hinduism*, Penguin Books, Harmondsworth, Middlesex UK, 1961.

329. Seymour, Jane, Hungry for a new revolution, *New Scientist*, 30 March 1996.

330. Shanks, Michael, *The Stagnant Society*, Penguin, Harmondsworth, Middlesex, UK, 1961.

331. Shapiro, Debbie, *Your Body Speaks Your Mind: Understand how your thoughts and emotions affect your health*, Piatkus, London, 1996.

332. Sheldrake, Rupert, *A New Science of Life*, Blond and Briggs, London, 1981.

333. Sheldrake, Rupert, *The Presence of the Past*, William Collins, London, 1988.

334. Sheldrake, Rupert, *Dogs That Know When Their Owners Are Coming Home: and other unexplained powers of animals*, Hutchinson, London 1999.

335. Sheldrake, Rupert, *The Sense of Being Stared At: And other aspects of the extended mind*, Hutchinson, London, 2003.

336. Shotyk, William, M. Krachler and B. Chen, *J. Environmental Monitoring*, **8**, 288, 2006.

337. Sicher, Fred, Elizabeth Targ, Dan Moore and Helene Smith, A randomized double-blind study of the effect of distant healing in a population with advanced AIDS, *Western Journal of Medicine*, **169** (6), 365-363, 1998.

338. Sigman, Aric, *Remotely Controlled: How television is damaging our lives and what we can do about it*, Ebury (Vermillion), London 2005.

339. Silf, Margaret, *The Way of Wisdom*, Lion Hudson, Oxford, 2006.

340. Simon, Julian L., *The Ultimate Resource*, Princeton University Press, Princeton, New Jersey, 1981.

341. Singer, J.A. and P. Salovey, *The Remembered Self: Emotion and memory in personality*, The Free Press, New York, 1993.

342. Sissons, Michael (Ed.), *A Countryside for All*, Vintage, London, 2001.

343. Smart, Ninian, *The Religious Experience of Mankind*, Charles Scribner's Sons, New York, 1969; William Collins (Fontana), London, 1971.

344. Smith, Adam, *The Theory of Moral Sentiments* 1759, in *The Works and Correspondence of Adam Smith*, Oxford University Press, Oxford, 1976, Vol. 1, Glasgow Edition.

345. Smith, Adam, *Enquiry into the Nature and Causes of the Wealth of Nations* 1776, in *The Works and Correspondence of Adam Smith*, Oxford University Press, Oxford, 1976, Vol. 2, Glasgow Edition.

346. Smith, John Maynard, *The Theory of Evolution*, Penguin, Harmondsworth, Middlesex, UK, 1958.

347. Society. Editorial from *The Independent*, 26 March, 1995.

348. Solomon, Grant and Jane Solomon, *The Scole Experiment: Scientific evidence for life after death*, Piatkus, London, 1999.

349. Solomon, P.R. *et al.* Ginkgo for memory enhancement: a randomized controlled trial, *J. Amer. Med. Assocn.*, **288**(7), 835-40, 2002.

350. Soros, George, 'Capital Crimes', *Atlantic Monthly*, January 1997.

351. Sprigge, T.L.S., *Theories of Existence*, Penguin, Harmondsworth, Middlesex, UK, 1984.

352. Spinoza, Baruch, *Tractatus Politicus* (Political Treatise), edited by A.G. Wernham, Oxford University Press, 1958.

353. Spinoza, Baruch, *Ethics*, 1677.

354. Stace, W.T., *Mysticism and Philosophy*, Macmillan, Basingstoke, Hampshire, UK, 1961.

355. Stanley, Michael W., *Emanuel Swedenborg: Essential readings*, Thorsons (Crucible), 1988.

356. Sternberg, Esther M. *The Balance Within: The science connecting health and the emotions*, W.H. Freeman, New York, 2001.

357. Stewart, Ian, *Nature's Numbers: Discovering order and pattern in the universe*, Weidenfeld and Nicholson, London, 1995.

358. Storr, Anthony, *Jung*, William Collins (Fontana), London, 1973.

359. Stray, Geoff, *Catastrophe or Ecstasy: A complete guide to end-of-time predictions*, Vital Signs Publishing, 2005.

360. Swimme, Brian, *The Hidden Heart of the Cosmos: Humanity and the New Story*, Orbis Books, Maryknoll, New York, 1996.

361. Taylor, Barbara Brown, *The Luminous Web*, Cowley Publications, Cambridge, MA, USA, 2000.

362. Taylor, R.J. *Food Additives*, Wiley, Chichester, UK, 1980.

363. Teachers. 'Does anyone ever fail teacher training?' by Phil Revel, *Guardian*, Tues 16 May 2006.

364. Teilhard de Chardin, Pierre, *The Phenomenon of Man*, William Collins, London, and Harper, New York, 1959; English trans. of *Le Phenoméne Humain*, 1955.

365. Thatcher, Margaret, in an interview reported in *Woman's Own* in 1987.

366. Thomas, David, A study on the mineral depletion of the foods available to us as a nation over the period 1940 to 1991, *Nutrition and*

Health, **17**, 85-115, 2003.

367. Toffler, Alvin, *Future Shock*, Bodley Head, London, 1970.

368. Toffler, Alvin, *The Third Wave*, William Collins, London, 1980.

369. Tompkins, Peter and Christopher Bird, *The Secret Life of Plants*, Penguin, Harmondsworth, Middlesex UK, 1973.

370. Trigg, Roger, *Rationality and Religion*, Blackwell, Oxford, UK, 1998.

371. Trigg, Roger, *Ideas of Human Nature*, Blackwell, Oxford, UK, 2nd edn. 1999.

372. Truss, Lynne, *Eats Shoots and Leaves*, Profile Books, London, 2003.

373. United Nations Committee on Economic, Social and Cultural Rights (CESCR), Twenty-second session in Geneva, 25 April-12 May 2000, General Comment #14 on the Right to Health.

374. Vaughan, Alan, *Incredible Coincidence*, Lippincott, New York, 1979.

375. Walton, John C., Organization and the origin of life, *Origins*, **4**(1), 16-35, 1977.

376. Ward, Keith, *Images of Eternity*, Darton, Longman and Todd, London, 1987; Oneworld, Oxford, UK, 1993.

377. Ward, Keith, *God, Chance and Necessity*, Oneworld, Oxford, UK, 1996.

378. Watson, Lyall, *The Nature of Things*, Hodder and Stoughton, London, 1990.

379. Watts, Alan W., *The Way of Zen*, Pantheon, New York, 1957; Thames and Hudson, London, 1957; Penguin, Harmondsworth, Middlesex, 1962.

380. Weber, Max, *Die protestantische Ethik und der Geist des Kapitalismus*, 1904/5; English trans. *The Protestant Ethic and the Spirit of Capitalism*, 1930.

381. Weber, Max, *Economy and Society*, Original German edn. 1921; trans. and ed. by Guenther Roth and Claus Wittich, Bedminster Press,

New York, 1968.

382. Weil, Simone, *L'Enracinement*, 1949; trans. by A.F. Wills, *The Need for Roots*, Routledge and Kegan Paul, London, 1952.

383. Weiss, Brian, *Many Lives, Many Masters*, Simon and Schuster, New York, 1988; Piatkus, London, 1994.

384. Wheeler, William Morton, *Ants, Their Structure, Development and Behavior*, Columbia University Press, 1910; The *Social Insects: Their origin and evolution*, Kegan Paul, London, 1928.

385. White Eagle, *The Quiet Mind*, The White Eagle Publishing Trust, Liss, Hampshire, UK, 1972.

386. White Eagle, *The Still Voice*, The White Eagle Publishing Trust, Liss, Hampshire, UK, 1981.

387. White, Ruth, *Karma and Reincarnation*, Piatkus, London, 2000.

388. Whitehead, Alfred North, *The Concept of Nature*, Cambridge University Press, Cambridge, UK, 1920.

389. Whitehead, Alfred North, *Process and Reality*, Macmillan, New York, 1929.

390. Wilber, Ken, *A Theory of Everything*: *An integral vision for business, politics, science and spirituality*, Gill and Macmillan (Gateway), Dublin, 2001, p.40.

391. Wilce, Hilary, *Help Your Child Succeed At School*, Piatkus, London, 2004.

392. Wilkinson, Richard, *Unhealthy Societies: The afflictions of inequality*, Routledge, London, 1996.

393. Willett, W.C. Balancing life-style and genomics research for disease prevention, *Science*, **296**, 695-98, 2002.

394. Williamson, Marianne, *The Gift of Change*, Harper-Collins, New York, 2004.

395. Wilson, Colin, *Afterlife*, Grafton/Collins, London, 1957.

396. Wilson, E.O. (ed.), *Biodiversity*, National Academy Press, Washington, D.C., 1988.

397. Witteveen, H.J., *Universal Sufism*, Element Books, Shaftesbury,

Dorset, UK, 1997.

398. Witteveen, H.J., *Sufism in Action*, Vega, London, 2003.

399. Woolger, Roger, *Other Lives, Other Selves*, Wrekin Trust, 1990.

400. Yellowlees, Walter, Ill Fares The Land, in *Soil, Food and Health*, Wholefood Trust, pp.15-42, 1989.

401. Ziman, John, Reliable Knowledge, Cambridge University Press, Cambridge, UK,

402. Zohar, Danah, and Ian Marshall, *Spiritual Intelligence,* Bloomsbury, London, 2000.

403. Zukav, Gary, *The Dancing Wu Li Masters: An overview of the new physics*, William Morrow, USA, 1979; Rider/Hutchinson, London, 1979.

404. Zukav, Gary, *The Seat of the Soul: An inspiring vision of humanity's spiritual destiny*, Rider, London, 1990.

INDEX